VIETNAM STUDIES

U.S. ARMY SPECIAL FORCES 1961–1971

by

Colonel Francis J. Kelly

DEPARTMENT OF THE ARMY

WASHINGTON, D.C., 2004

Library of Congress Catalog Card Number: 72:600263

First Printed 1973—CMH Pub 90–23–1

For sale by the Superintendent of Documents, U.S. Government Printing Office
Internet: bookstore.gpo.gov Phone: toll free (866) 512–1800; DC area (202) 512–1800
Fax: (202) 512–2250 Mail: Stop SSOP, Washington, DC 20402–0001

Foreword

The United States Army has met an unusually complex challenge in Southeast Asia. In conjunction with the other services, the Army has fought in support of a national policy of assisting an emerging nation to develop governmental processes of its own choosing, free of outside coercion. In addition to the usual problems of waging armed conflict, the assignment in Southeast Asia has required superimposing the immensely sophisticated tasks of a modern army upon an underdeveloped environment and adapting them to demands covering a wide spectrum. These involved helping to fulfill the basic needs of an agrarian population, dealing with the frustrations of antiguerrilla operations, and conducting conventional campaigns against well-trained and determined regular units.

As this assignment nears an end, the U.S. Army must prepare for other challenges that may lie ahead. While cognizant that history never repeats itself exactly and that no army ever profited from trying to meet a new challenge in terms of the old one, the Army nevertheless stands to benefit immensely from a study of its experience, its shortcomings no less than its achievements.

Aware that some years must elapse before the official histories will provide a detailed and objective analysis of the experience in Southeast Asia, we have sought a forum whereby some of the more salient aspects of that experience can be made available now. At the request of the Chief of Staff, a representative group of senior officers who served in important posts in Vietnam and who still carry a heavy burden of day-to-day responsibilities has prepared a series of monographs. These studies should be of great value in helping the Army develop future operational concepts while at the same time contributing to the historical record and providing the American public with an interim report on the performance of men and officers who have responded, as others have through our history, to exacting and trying demands.

All monographs in the series are based primarily on official records, with additional material from published and unpublished secondary works, from debriefing reports and interviews with key participants, and from the personal experience of the author. To facilitate security clearance, annotation and detailed bibliography have been omitted from the published version; a fully documented account with bibliography is filed with the Office of the Chief of Military History.

Colonel Francis John Kelly is eminently qualified to write the story of U.S. Army Special Forces. In 1960 he chaired the committee at the Command and General Staff College which produced the U.S. Army's first definitive approach to counterinsurgency, "The Role of the U.S. Army in the Cold War." He also wrote and conducted the Senior Officer Counterinsurgency Program course of study at the U.S. Army War College and served as a division chief in the Special Warfare Directorate, Office of the Deputy Chief of Staff for Military Operations, Department of the Army. For two years he commanded the 1st Special Forces Group (Airborne) on Okinawa, which provided multiple operational teams for combat service in Vietnam. From June 1966 to June 1967 he commanded the 5th Special Forces Group (Airborne) in Vietnam. Upon his return, he became the Commander of the Combat Development Command Institute for Strategic and Stability Operations at Fort Bragg. In all these positions, he strongly influenced the development of tactics and techniques, equipment, organization, and doctrine. After service in Vietnam, Colonel Kelly undertook the task of complete reorganization of the basic unit, the Special Forces Group, at the same time revising the doctrine. In September 1970 he was assigned as Senior Army Advisor to the State of Colorado in Denver.

VERNE L. BOWERS
The Adjutant General
Major General, USA

15 September 1972
Washington, D.C.

Preface

As long ago as 1957, U.S. Army Special Forces soldiers were in the Republic of Vietnam, going about their business of training, advising, and assisting members of the Vietnamese Army. Despite the old Army witticism about never volunteering for anything, the Special Forces soldier is, in fact, a double volunteer, having first volunteered for airborne training and then again for Special Forces training. From a very meager beginning but sustained by a strong motivation and confidence in his mission, the Special Forces soldier has marched through the Vietnam struggle in superb fashion.

In 1957 some fifty-eight Vietnamese soldiers were given military training by Special Forces troops. Ten years later the Special Forces were advising and assisting over 40,000 paramilitary troops, along with another 40,000 Regional Forces and Popular Forces soldiers. This monograph traces the development and notes the progress, problems, successes, and failures of a unique program undertaken by the U.S. Army for the first time in its history. It is hoped that all the significant lessons learned have been recorded and the many pitfalls of such a program uncovered. I am indebted to Major James M. Scott, Corps of Engineers, for his assistance on the Engineer effort. I am responsible for the conclusions reached, yet my thought processes could not escape the influence of the many outstanding officers and men in the Special Forces who joined in the struggle. Particularly, I must take note of the contributions of the Special Forces noncommissioned officers, without question the most competent soldiers in the world.

With the withdrawal of the Special Forces from Vietnam in 1971, the Army could honestly lay claim to a new dimension in ground warfare—the organized employment of a paramilitary force in sustained combat against a determined enemy. I know I speak for my predecessors and successors in claiming that the 5th Special Forces Group (Airborne) was the finest collection of professional soldiers ever assembled by the U.S. Army, anywhere, anytime.

Washington, D.C.
15 September 1972

FRANCIS JOHN KELLY
Colonel, Armor

Contents

PART ONE

The Early Years: 1961–1965

Chapter *Page*

I. INTRODUCTION 3

The Origin of the Special Forces 3
The Unconventional Requirements 6

II. BEGINNINGS OF THE CIVILIAN IRREGULAR DEFENSE GROUP PROGRAM 19

The Montagnard Culture 20
The Buon Enao Experiment 24
Command and Control During the Buon Enao Period 28
Early Paramilitary Programs 32
Operation Switchback, November 1962–July 1963 35
The Turnover of Buon Enao 41

III. THE CIDG PROGRAM UNDER THE U.S. MILITARY ASSISTANCE COMMAND, VIETNAM: JULY 1963–MAY 1965 45

Combat Operations 49
Logistics 57
Civic Action and Psychological Operations 59
The Montagnard Uprising 63
The 5th Special Forces Group (Airborne) 64

PART TWO

The Middle Years: 1965–1968

IV. THE CIDG PROGRAM BEGINS TO MATURE 77

Chapter | Page

V. THE COMBINED SPECIAL FORCES CIDG ON THE OFFENSIVE 96

Intelligence Operations Overhauled 99
The Tet Offensive: The CIDG Grows Up 122

VI. UNCONVENTIONAL OPERATIONS 134

PART THREE

The Latter Years: 1968-1971

VII. THE VIETNAMESE ARMY ABSORBS THE CIDG PROGRAM 151

Summary by Corps of Combat Activity During 1970 154
The Stand-Down 155

VIII. THE BALANCE SHEET 160

Credits 168
Debits 171
The Future of the Special Forces 173

Appendix

A. Chronology of U.S. Army Special Forces, 15 April 1970–1 March 1971 177
B. U.S. Army Special Forces Honors 180
C. List of Special Forces Camps, 1961–1971 182
D. U.S. Army Special Forces Engineer Activities 194
E. After Action Report MIKE Force: ATTLEBORO, 1–7 November 1966 199
F. After Action Report: Operation BLACKJACK 33 204
G. Chronology of Logistic Events 214

GLOSSARY 216

INDEX 221

Tables

No.		*Page*
1.	Disposition of U.S. Special Forces Detachments, December 1962	38
2.	Ethnic Background of Strike Forces at U.S. Special Forces CIDG Camps, October 1964	50
3.	Organization of U.S. Special Forces Detachments	65
4.	Status Report, U.S. Special Forces CIDG Camps, October 1964	66
5.	U.S. Special Forces CIDG Camps Established in Vietnam, July 1961–October 1964	68
6.	Expansion of 5th Special Forces Group and Span of Advisory Operations, October 1964–July 1967	82
7.	Comparative Data on Combat Operations, 1966–1967	104
8.	Conversion of U.S. Special Forces CIDG Camps to Vietnam Army Ranger Camps, 27 August 1970–15 January 1971	152

Charts

1.	Special Forces Group (Airborne)	10
2.	Headquarters and Headquarters Company, Special Forces Group (Airborne)	11
3.	Special Forces Company, Special Forces Group (Airborne)	12
4.	Signal Company, Special Forces Group (Airborne)	13
5.	Organization of Civilian Irregular Defense Group Company in 1964	48
6.	Organization of Mobile Strike Force Company	117
7.	Command Structure of 5th Special Forces Group (Airborne)	123
8.	Mobile Strike Force Organization	136
9.	Organization of Project Delta	138
10.	Organization of Projects Sigma and Omega	139

Maps

1.	Distribution of Major Ethnic Groups, 1964	21
2.	Buon Enao Expansion	29
3.	U.S. Special Forces Deployment, 15 October 1962	31
4.	U.S. Special Forces Deployment, 1 June 1963	39
5.	U.S. Special Forces Deployment, 20 October 1964	51
6.	Location of 5th Special Forces Detachments, 31 August 1967	103

Illustrations

	Page
Vietnamese Soldiers and Special Adviser on Training Mission	5
Daily Medical Visit to a Montagnard Hamlet	8
Village Self-Defense	14
Strike Force Billet at Vietnamese Special Forces Camp	16
Rhade Village of Buon Enao	23
U.S. Special Forces Soldier Removes Jammed Bazooka Shell	33
Helicopter Lifts Off After Dropping Supplies to Isolated Special Forces Team	36
U.S. Special Forces Camp	47
Treating a Child at the Special Forces Dispensary	60
Medical Man Giving Dapsone Pills to a Leper	61
Montagnards and Special Forces Soldiers at Ease	78
Strike Force Assault Troops Leaving Helicopter	79
'Special Forces Men and Civilian Irregulars on Patrol Pause To Eat	86
Special Forces Soldiers Between Missions	89
Soldiers Firing an 81-mm. Mortar	91
Diagram of a Fighting Camp	106
Vietnamese Strike Force Troops Man Airboat in River Operation	107
Last Minute Briefing of Strike Force Before Attack	109
Candy After Medicine	120
Teaching Wife of Vietnamese Soldier To Treat His Foot Wound	125
Civilian Irregulars Learning To Fire a Machine Gun	128
Special Forces Soldier Directs Emergency Removal of Troops	146
Vietnamese Special Forces Move Up River on a Mission	162
Training Jump	167
Men of a Special Forces A Team	170

All illustrations are from the Department of Defense files.

OPPRESSO LIBER

PART ONE

THE EARLY YEARS: 1961–1965

CHAPTER I

Introduction

The same events and pressures that shaped directly or indirectly the major part of American foreign policy during the last twenty years led to the formation and activation of the U.S. Army Special Forces.

In February of 1950 the United States recognized a quasi-independent Vietnam within the French Union and first began to consider granting aid to the French forces fighting against Communist insurgency in Indochina. In May of the same year the United States agreed to grant military and economic aid. American involvement in post-World War II Southeast Asia had begun. Four years later, in May 1954, the French Army was defeated by the Viet Minh—the Communist-supported Vietnam Independence League—at Dien Bien Phu, and under the Geneva armistice agreement Vietnam was divided into North and South Vietnam. In the course of those four years the policy-makers of the United States had an opportunity to observe the struggle of France with the insurgents and to become familiar with the political and military situation in Vietnam. It was also during those years that the U.S. Army Special Forces came into existence.

Origin of the Special Forces

The 1st Special Service Force of World War II is considered the antecedent of the present U.S. Army Special Forces. In the spring of 1942 the British Chief of Combined Operations, Vice Admiral Lord Louis Mountbatten, introduced to U.S. Army Chief of Staff General George C. Marshall a project conceived by an English civilian, Geoffrey N. Pike, for the development of special equipment to be used in snow-covered mountain terrain. This plan, named PLOUGH, was designed for attack on such critical points as the hydroelectric plants in Norway upon which the Germans depended for mining valuable ores. American manufacturers working on equipment for the project developed a tracked vehicle known as the Weasel and eventually standardized as the M29.

General Marshall concluded that an elite force recruited in

Canada and the United States would be the best military organization for conducting the raids and strikes; he selected an American, Lieutenant Colonel Robert Tryon Frederick, to assemble, organize, train, and command the U.S.–Canadian 1st Special Service Force.

Made up of three regiments of two battalions each, the unit became a separate branch of the service, with the crossed arrows of the Indian Scouts, by then inactivated, as its insignia. The men were trained in demolitions, rock-climbing, amphibious assault, and ski techniques, and were given basic airborne instruction. They fought under Allied command with great bravery and considerable success in the Aleutians, North Africa, Italy, and southern France. The 1st Special Service Force got its nickname, "The Devil's Brigade," during the Italian campaign from a passage in the captured diary of a dead German officer who had written: "The black devils are all around us every time we come into line and we never hear them." The force was inactivated in southern France near the end of World War II.

On 20 June 1952 the first of the Special Forces groups, the 10th Special Forces Group, was activated at Fort Bragg, North Carolina; it became the nucleus of the Special Warfare Center, now known as the John F. Kennedy Center for Military Assistance, at Fort Bragg. The next unit to be formed was the 77th Special Forces Group, which was also activated at Fort Bragg, on 25 September 1953.

By July 1954 the U.S. Military Assistance Advisory Group, Vietnam, numbered 342. In October of that year President Dwight D. Eisenhower promised direct aid to the government of South Vietnam, headed at that time by Premier Ngo Dinh Diem. From 1954 to 1956 Viet Minh cadres were forming action committees to spread propaganda and to organize the South Vietnamese to oppose their own government. In July 1955 the People's Republic of China announced an agreement to aid the Viet Minh, and the Soviet Union announced aid to Hanoi. In August Diem's government rejected for the third time Hanoi's demands for general elections throughout the two Vietnams, and in October South Vietnam was proclaimed a republic by Premier Diem, who became the first president.

U.S. Special Forces troops actually worked in Vietnam for the first time in 1957. On 24 June 1957 the 1st Special Forces Group was activated on Okinawa, and in the course of the year a team from this unit trained fifty-eight men of the Vietnamese Army at the Commando Training Center in Nha Trang. The trainees would later become the nucleus, as instructors and cadre, for the first Vietnamese Special Forces units.

VIETNAMESE SOLDIERS AND SPECIAL FORCES ADVISER *on training mission.*

In 1959 and 1960 the insurgents in South Vietnam, known to the South Vietnamese as Viet Cong, a contraction for Vietnamese Communists, grew in number and in power to terrorize the people. Clashes between government forces and armed Viet Cong increased in number from 180 in January 1960 to 545 in September of that year. Thirty Special Forces instructors were sent from Fort Bragg to South Vietnam in May 1960 to set up a training program for the Vietnamese Army.

President John F. Kennedy announced on 21 September 1961 a program to provide additional military and economic aid to Vietnam. The government of the United States was by this time deeply concerned over the insurgency in South Vietnam and the necessary steps were being taken to help the republic to deal with it.

On 21 September 1961 the 5th Special Forces Group, 1st Special Forces, which would eventually be charged with the conduct of all Special Forces operations in Vietnam, was activated at Fort Bragg. It was at this point, in the fall of 1961, that President Kennedy began to display particular interest in the Special Forces. His enthusiasm, based on his conviction that the Special Forces had great potential as a counterinsurgency force, led him to become a very powerful advocate for the development of the Special Forces program within the Army. President Kennedy himself made a visit to the Special Warfare Center in the fall of 1961 to review the program, and it was by his authorization that Special Forces troops were allowed to wear the distinctive headgear that became the symbol of the Special Forces, the Green Beret.

Up to 1961 the government of South Vietnam and the U.S. Mission in Saigon in dealing with the insurgency had placed primary emphasis on developing the regular military forces, which for the most part excluded the ethnic and religious minority groups. Under the sponsorship of the U.S. Mission in Saigon, however, several programs were initiated in late 1961 to broaden the counterinsurgency effort by developing the paramilitary potential of certain of these minority groups. Special Forces detachments were assigned to the U.S. Mission in Saigon to provide training and advisory assistance in the conduct of these programs, which eventually came to be known collectively as the Civilian Irregular Defense Group (CIDG) program. The development of paramilitary forces among the minority groups became the primary mission of the Special Forces in Vietnam.

Originally attention was concentrated on the Montagnards, who lived in the strategic Central Highlands. The first step was taken in October 1961 with the beginning of a project designed to prevent the Rhade tribesmen in Darlac Province from succumbing to Viet Cong control. Exploratory talks were held with Rhade leaders in Darlac to seek their participation in a village self-defense program. One Special Forces medical noncommissioned officer participated in that first effort.

Early in 1962 the government of the United States under President Kennedy began to set up the actual interdepartmental machinery for aiding South Vietnam. The Executive Branch, the Department of State, the Department of Defense, the Joint Chiefs of Staff, the United States Information Agency, the Agency for International Development, and the Central Intelligence Agency were all involved. Because of the nature of the growing conflict in Vietnam and because the Special Forces was designed for unconventional warfare, it was inevitable that the Special Forces would play a conspicuous role. It was also plain that the actions and suggestions of the various government agencies would heavily influence that role.

The Unconventional Requirements

In 1961 a serious examination of the responsibility of the U.S. Army in the cold war had been instituted at the Command and General Staff College, Fort Leavenworth, Kansas. The strategy of wars of liberation" as practiced by the Communists was analyzed in detail, lessons learned were reviewed, and a comprehensive assessment of U.S. Army capabilities was prepared to show the resources available to the United States for resisting insurgency. Doctrinal gaps were identified, mission statements amended, and training requirements defined.

The initial efforts of the United States to counter subversive insurgency in Vietnam quickly became a co-ordinated departmental endeavor at the highest national level. In addition to mustering the talent, technical ability, and equipment of the military, the government called on each department to nominate certain units and numbers of forces which it considered best prepared to deal with the peculiarities of countering insurgencies. The U.S. Army chose as its vanguard unit the Special Forces, whose highly trained group of combat specialists numbered at the time approximately 2,000 men.

An assessment of insurgent strategy, particularly as it was being practiced at the time in the Republic of Vietnam, indicated that good use could be made there of the U.S. Army Special Forces. The requirement for a unit that was combat-oriented, capable of performing with relative independence in the field, ruggedly trained for guerrilla operations, and geared for co-operation with the Vietnamese was admirably met in the organization, training, equipment, and operational procedures of the U.S. Army Special Forces.

In November 1961 the first medical specialist troops of the Special Forces were employed in Vietnam in a project originally designed to provide assistance to the Montagnard tribes in the high-plateau country around Pleiku. Out of this modest beginning grew one of the most successful programs for using civilian forces ever devised by a military force—the Civilian Irregular Defense Group. Eventually the organization, development, and operation of the Civilian Irregular Defense Group proved to be the chief work of the U.S. Special Forces in the Vietnam War.

Despite the size and complexity of the program, however, the U.S. Special Forces participated in a number of other activities in the course of their stay in Vietnam, including training, advisory, and operational missions. Any comprehensive story of what the Special Forces did in Vietnam must include some account of these missions. The nature, scope, and success of the Civilian Irregular Defense Group program will nevertheless occupy a substantial part of this study.

U.S. Special Forces occupied a somewhat unusual position *vis-à-vis* the Vietnamese Army, the Vietnamese Special Forces, and the

DAILY MEDICAL VISIT TO A MONTAGNARD HAMLET

indigenous population involved in the program. The rules of engagement specified that in most instances the U.S. Special Forces would serve, technically at least, in an advisory capacity to the Vietnamese Special Forces, which was charged with the direct command responsibility for the Civilian Irregular Defense Group. There were exceptions to this. For instance, the troops known as the mobile guerrilla forces were originally commanded and controlled directly by soldiers of the U.S. Special Forces. For the most part, however, the Vietnamese were in command; the Americans were there to assist them—not to assume any command. In practice, as will be seen, this arrangement was not firmly and universally adhered to from the start. There were degrees of compliance that varied considerably from one case to the next. Many of the early problems encountered by the Civilian Irregular Defense Group came from the U.S. Special Forces–Vietnamese Special Forces command and control structure imposed upon it. The obvious dilemma of two command figures, each with his own judgments, arose. No less a factor, especially in the years 1962 and 1963, was the mutual mistrust and dislike between the civilian irregulars, especially the Montagnards, and the Vietnamese military men who were commanding them.

The U.S. Special Forces had been created by the Army for the purpose of waging unconventional warfare, which by 1964 was defined in the Dictionary of United States Army Terms as "The three inter-related fields of guerrilla warfare, evasion and escape, and subversion against hostile states. Unconventional warfare, operations," the dictionary stated, "are conducted within enemy or enemy-controlled territory by predominantly indigenous personnel, usually supported and directed in varying degrees by an external source."

The Special Forces was defined in Field Manual 31–21, Special Forces Operations, in terms of its role, mission, and capabilities. Its role was to assume any responsibility and carry out any mission assigned to it by the Army. Its missions were many and varied because of the Special Forces' organization, flexible command arrangements, tailored logistical and fiscal procedures, and highly trained men. Chief among them were planning, conducting, and supporting unconventional warfare and internal security, or "stability" operations. Special Forces troops were capable of training, advising, and providing operational, logistical, and fiscal support for foreign military or paramilitary forces. They were able to infiltrate by air, land, or water, sometimes penetrating deep into enemy territory for the purpose of attacking strategic targets, rescuing friendly troops, or collecting intelligence. Special Forces troops also trained other American and allied forces in Special Forces techniques. To a large extent these definitions were determined by the problems that faced the Army and how the Army used the Special Forces to solve them. The Special Forces units evolved in response to the demands placed upon them.

The basic structure of the Special Forces Group (Airborne) consisted of a headquarters and headquarters company, three or more line Special Forces companies, a signal company, and an aviation detachment. (*Chart 1*) The headquarters and headquarters company encompassed all the usual staff sections for command and control, as well as the major portion of the group medical capability and the parachute rigging and air delivery elements. (*Chart 2*) The line Special Forces company was commanded by a lieutenant colonel and was normally composed of an administrative detachment and an operations detachment C, which commanded three operations detachment B's, each of which commanded four operations detachment A's. The A detachment was the basic twelve-

CHART 1—SPECIAL FORCES GROUP (AIRBORNE)

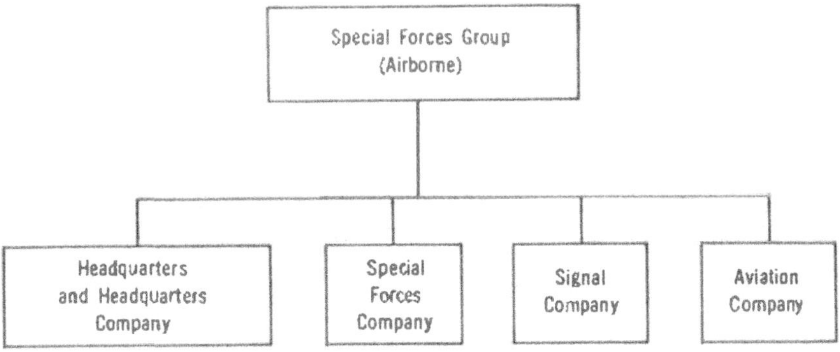

man unit of the Special Forces. (*Chart 3*) Supporting the entire group with communications was the signal company, which, in terms of personnel, technical equipment, and communications capabilities, resembled a battalion more than it did the usual signal company. (*Chart 4*)

In the early years of Special Forces involvement in Vietnam, 1961–1965, the concept of how best to employ the forces was developed, put into practice, and adjusted empirically. The government of the United States and the government of South Vietnam were dealing with a Communist-inspired insurgency, and for the United States it was a new experience. Many local tactics were attempted on a "let's-try-it-and-see-what-happens" basis. If something worked, then it became an acceptable counterinsurgency tactic; if it did not, it was dropped.

During these formative years, it became clear that the part the U.S. Special Forces was to play would differ from the role foreseen for it when it was created in the 1950s. At that time, the troops of the force as organized were capable of waging unconventional war under conventional war conditions. The war in Vietnam, however, never fell smoothly into the conventional category. In Vietnam "enemy or enemy-controlled territory" was the countryside of South Vietnam, the government of which had invited U.S. military presence. The enemy insurgents were guerrillas themselves. Instead of waging guerrilla warfare against conventional forces in enemy territory, the U.S. Special Forces troops were to find themselves attempting to thwart guerrilla insurgency in "friendly" territory.

At first the Civilian Irregular Defense Group program was concerned with what was called area development. The goal was to

Chart 2—Headquarters and Headquarters Company, Special Forces Group (Airborne)

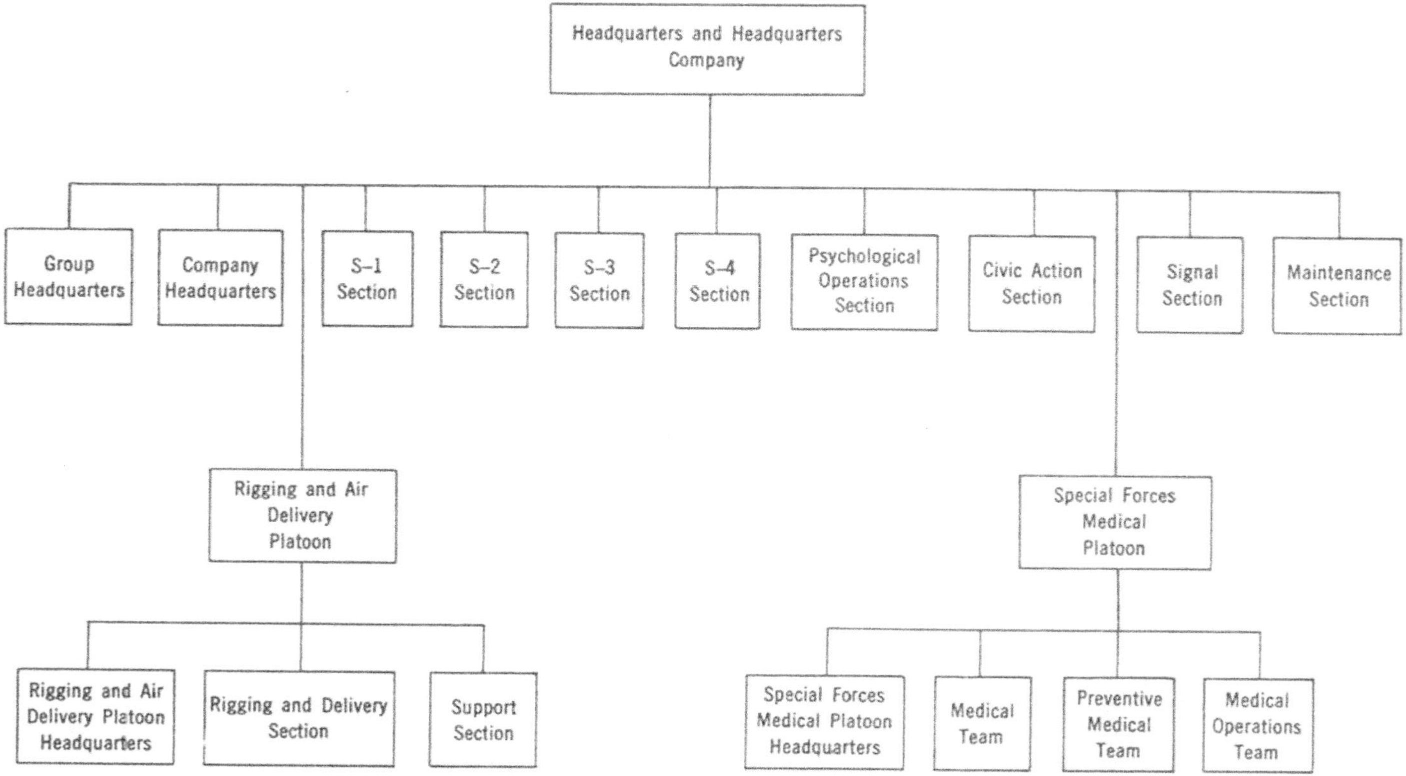

CHART 3—SPECIAL FORCES COMPANY, SPECIAL FORCES GROUP (AIRBORNE)

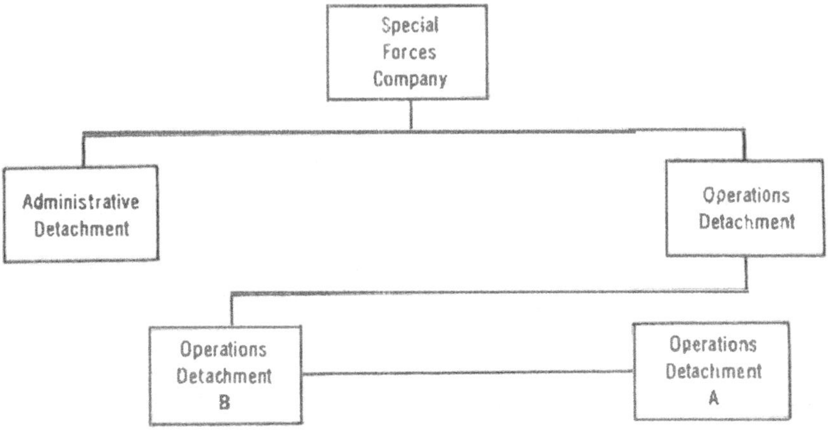

provide an area with security from Viet Cong influence and terror, to help the people develop their own self-defense program, and, if possible, to enlist support for the government of Vietnam from its own citizens. Operations took an offensive turn only because many of the areas involved were already effectively controlled by the Viet Cong.

In late 1960 the response of the governments of Vietnam and the United States, whose military involvement at that time consisted of the presence of a Military Assistance Advisory Group, to the mounting Communist insurgency was to increase the size and effectiveness of Vietnam's conventional military forces. For the most part, these did not include the ethnic and religious minority groups in the highlands of the central and northern portions of South Vietnam and in the rural lowlands of the Mekong Delta. Under the sponsorship of the U.S. Mission in Saigon several programs were initiated in late 1961 to keep these minority groups from falling under the control of the Viet Cong. U.S. Special Forces detachments were assigned to the U.S. Mission to provide training and advice for the programs, the first of which was among the Montagnards.

Based primarily on the success of a pilot project involving the Rhade tribe around the village of Buon Enao in Darlac Province, the principal program centered on establishing area development centers in remote areas where there was little government control. The area development centers were bases of operation at which Special Forces detachments, working through Vietnamese Special Forces counterparts, assisted in the establishment of village defense

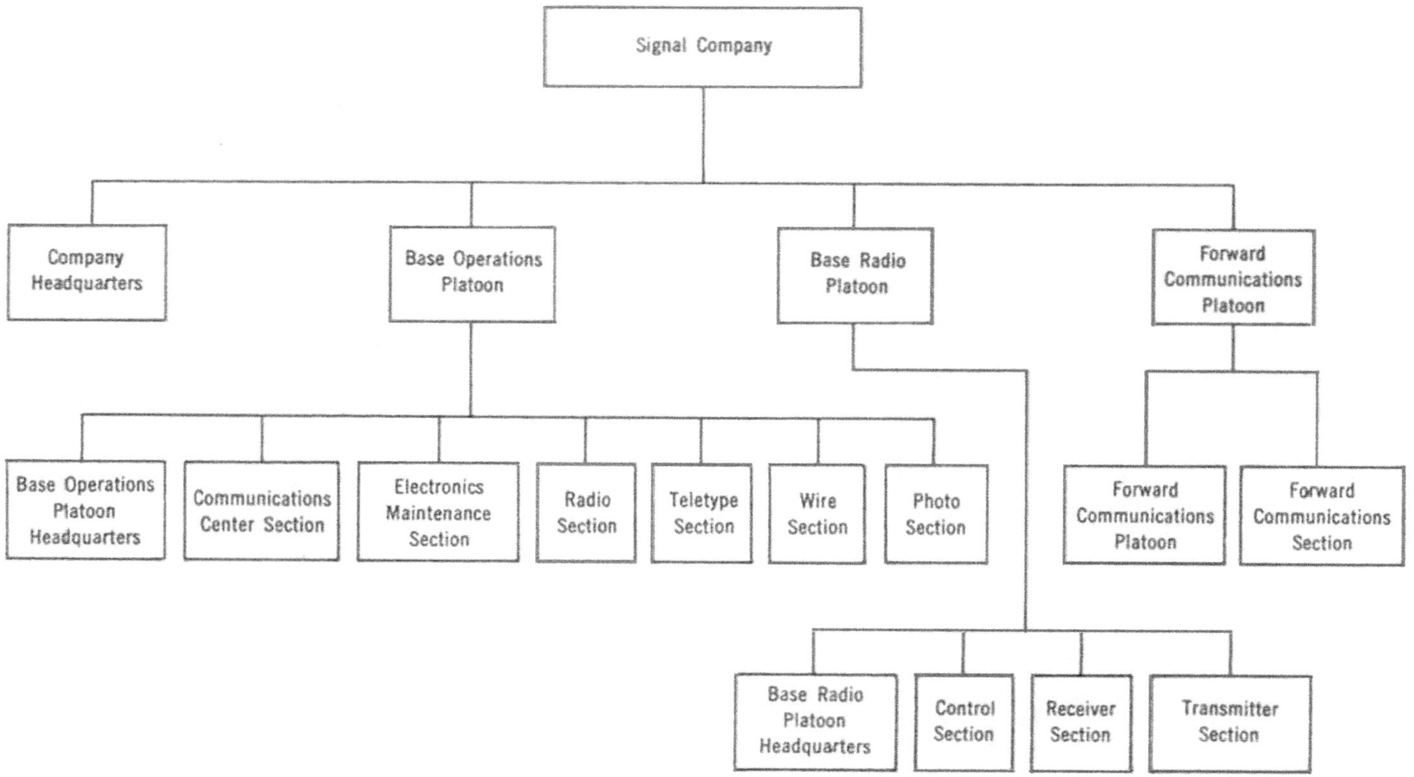
Signal Company
Company Headquarters
Base Operations Platoon
Base Radio Platoon
Forward Communications Platoon
Base Operations Platoon Headquarters
Communications Center Section
Electronics Maintenance Section
Radio Section
Teletype Section
Wire Section
Photo Section
Forward Communications Platoon
Forward Communications Section
Base Radio Platoon Headquarters
Control Section
Receiver Section
Transmitter Section

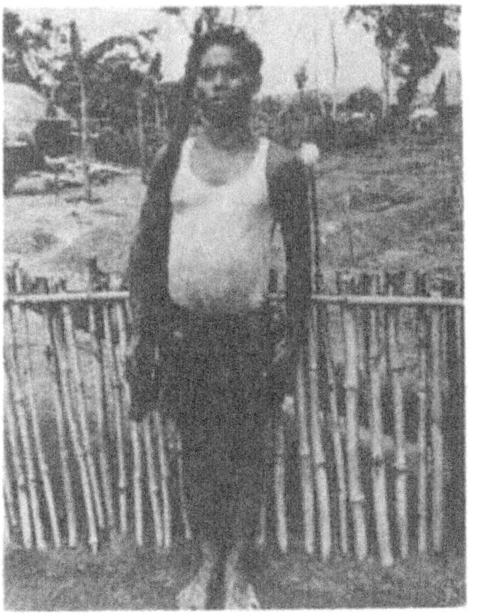

VILLAGE SELF-DEFENSE. *Moat with bamboo stakes and barbed wire fences surround some villages (left); Montagnard stands guard at his village (right).*

systems based on elementary training in small arms and mortars, with minimum tactics designed for squads and with occasional platoon maneuvers. The purpose of the program was to extend government control into areas where it was lacking and to generate in the local populace a more favorable attitude toward the government. It should be clearly understood that the United States initiated this program and encouraged it. The Vietnam government participated by employing the Vietnamese Special Forces, but the program was essentially an American project. In the beginning the local Vietnamese province-sector officials were less than enthusiastic.

In 1963 the area development program expanded toward the western borders of Vietnam. In 1964 the Civilian Irregular Defense Group assumed other missions calling for operations against Viet Cong war zones or so-called safe havens and the interdiction of Viet Cong infiltration routes in Vietnam. The Special Forces continuing commitment in terms of men involved in the CIDG program grew from one medical noncommissioned officer at Buon Enao in October 1961 to the 5th Special Forces Group (Airborne), numbering over 1,200 in October 1964.

In terms of program management and control, the early years can be divided into three periods: from November 1961 to November 1962 when the U.S. Mission was responsible for the Civilian

Irregular Defense Group program; from September 1962 to July 1963 during which responsibility for operations was gradually turned over to the U.S. Military Assistance Command, Vietnam, and the Army; and, finally, from July 1963 to the spring of 1965, when the conventional U.S. buildup began during which the Military Assistance Command, Vietnam, bore full responsibility for the Civilian Irregular Defense Group program. Throughout the early years, the Special Forces effort with civilian irregulars was characterized by rapid expansion, was dispersed over a wide area, and was subject to changing emphasis in missions. The program developed along largely unplanned lines in response to changing needs and opportunities.

From 1961 to 1965 more than eighty CIDG camps or area development centers were established. Many were built from the ground up (and down) in areas where the government had no effective control. Each camp was a self-contained and comprehensive counterinsurgency effort. U.S. Special Forces men provided advice and assistance in all aspects of camp administration and operations throughout each project site's existence, from initiation to turnover of the camp and its paramilitary assets to local Vietnamese authorities.

When the U.S. Military Assistance Command, Vietnam, began to assume responsibility for the Civilian Irregular Defense Group program in the fall of 1962, Special Forces detachments made the first assessments of areas in the selection of proposed campsites. Security was the prime consideration when the irregulars arrived at a new site. Often security forces from established camps were brought in until local forces could be recruited and trained.

Camp security occupied a major portion of the Special Forces detachment's time and effort. Few fortified camps were built in the early part of the program, but as it evolved the new camps were placed in "hot areas" and therefore required much more attention in both defense and security. Throughout the period, the Viet Cong harassed campsites and attacked several in reinforced battalion strength, with occasional success. After the successful attack on the camp at Hiep Hoa in November 1963, more emphasis was placed on making the camps strongly fortified positions.

One of the primary missions of Special Forces men at a camp was to advise and assist in the training of paramilitary forces recruited in that area. The Special Forces training program generally concentrated on strike force troops, although the Special Forces did participate in the training of hamlet militia, mountain scouts, and other irregular forces. The main problem in training civilian

STRIKE FORCE BILLET AT VIETNAMESE SPECIAL FORCES CAMP *amidst mortar and machine gun fortifications.*

irregular troops was establishing the respective roles of U.S. Special Forces and Vietnamese Special Forces. Theoretically, all training was a Vietnamese Special Forces responsibility, but most Vietnamese detachments were either unwilling or unable to undertake it.

Strike force operations consisted for the most part of patrols. Hundreds of contacts with the enemy occurred, and many small actions were fought. There was also a fair number of joint operations with regular Vietnam Army and Regional Forces units, particularly in 1964. In most operations, the major hindrance to success was the lack of accurate and timely intelligence.

The U.S. Special Forces men, aware of the importance of gathering intelligence, tried to emphasize that aspect of their missions and to set up intelligence nets that would produce information on the location of Viet Cong units and members of the local Viet Cong political organization. At the beginning of the program, there was no standing operating procedure for the procurement of intelligence. Each Special Forces detachment commander found it necessary to make working arrangements with his Vietnamese

Special Forces counterpart with regard to intelligence. Even after an agreement was finally reached in the spring of 1964, the Vietnamese Special Forces units were slow to accept U.S. Special Forces participation in intelligence operations. The language barrier proved to be a major obstacle to the U.S. Army in recruiting agents and acquiring information.

Perhaps the major problem encountered by U.S. Special Forces men in carrying out their mission with the civilian irregulars was their relationship with their Vietnamese counterparts. From the beginning of the program the role of the U.S. Special Forces detachment commander was to have been strictly advisory. All important responsibilities were to be assumed by the Vietnamese Special Forces, but unfortunately these were rarely shouldered by the Vietnamese Special Forces alone. To complicate the problem there were two vertical chains of command, with appropriate levels of horizontal counterpart co-ordination required up through the two commands. U.S. Special Forces men at this time, moreover, had received little training or indoctrination on what to expect from their Vietnamese counterparts, how to get along with them, and how to accomplish the operational mission through them.

The logistics involved in administering and resupplying the widely dispersed camps required unorthodox requisitioning and procurement procedures. The command and control structure up until May 1964 was unique because the nature of the Civilian Irregular Defense Group program demanded it. When conventional forces worked in conjunction with civilian irregular forces, however, this unconventional structure placed an exceptional burden of co-ordination on the Special Forces.

Counterguerrilla operations by strike force units were only a part of the counterinsurgency program at Civilian Irregular Defense Group sites. Civic action and psychological operations were also conducted as part of the Special Forces mission. Their objective was to raise the living standard of the people, to develop their identity with and their loyalty to the government, and to enlist their active support in defeating insurgents. The work of the detachment medical men was a major contribution to this effort. Throughout the period, however, these programs were hampered by the inability of Civilian Irregular Defense Group and other security forces to provide adequate protection to the local population against Viet Cong attacks and terrorism, poorly motivated local government representatives, and the lack of professionally qualified U.S. soldiers who knew the area to augment the Special Forces detachment for its civic action and psychological operations mission. In spite of these problems, Special Forces men on their own initiative accomplished many worthwhile civic action projects in this period. Emphasis is usually placed on the role the Special Forces played as soldiers in Vietnam. They were soldiers and good ones. But they were more than soldiers; they were, in a way, community developers in uniform too. The civic action accomplishments of the Special Forces are as much a source of pride to them as their accomplishments in the military arena, and justifiably so.

CHAPTER II

Beginnings of the Civilian Irregular Defense Group Program

There were two principal reasons for the creation of the Civilian Irregular Defense Group program. One was that the U.S. Mission in Saigon believed that a paramilitary force should be developed from the minority groups of South Vietnam in order to strengthen and broaden the counterinsurgency effort of the Vietnamese government. The other was that the Montagnards and other minority groups were prime targets for Communist propaganda, partly because of their dissatisfaction with the Vietnamese government, and it was important to prevent the Viet Cong from recruiting them and taking complete control of their large and strategic land holdings.

One major study of the situation in Southeast Asia concluded that in 1961 the danger of Viet Cong domination of the entire highlands of South Vietnam was very real, that the efforts of the Vietnamese Army to secure the highlands against Viet Cong infiltration were ineffective, and that the natural buffer zone presented by the highland geography and Montagnard population was not being utilized properly to prevent Communist exploitation. The government was, in fact, failing to exercise any sovereignty over its highland frontiers or its remote lowland districts in the Mekong Delta where other ethnic and religious minority groups were established. This lack of control deprived the government of any early intelligence of enemy attacks and any real estimate of Viet Cong infiltration. The Communists, on the other hand, continued to exploit the buffer zone, and there was always the danger that the insurgents would use this territory as a springboard into the more heavily populated areas.

The Vietnamese had not only made no attempt to gain the support of the Montagnards and other minority groups but in the past had actually antagonized them. Before 1954 very few Vietnamese lived in the highlands. In that year some 80,000 refugees from North Vietnam were resettled in the Montagnard area, and inevitably friction developed. Dissatisfaction among the Montagnards reached a point where in 1958 one of the principal tribes, the Rhade, organized a passive march in protest. Vietnamese officials countered by confiscating the tribesmen's crossbows and spears, an act that further alienated the Montagnards.

The indifference of the Vietnamese to the needs and feelings of the tribesmen grew directly out of their attitude toward the Montagnards, whom the Vietnamese had traditionally regarded as an inferior people, calling them "*moi*," or savages, and begrudging them their tribal lands. This attitude on the part of the Vietnamese plagued the Civilian Irregular Defense Group program from the beginning. Not until 1966 did the Vietnamese, in their desire to bring the tribes under government control, begin to refer to the Montagnards as *Dong Bao Thuong*, "compatriots of the highlands." Even so, the animosity between Montagnards and Vietnamese continued to be a major problem.

The Montagnard Culture

The Montagnards constitute one of the largest minority groups in Vietnam. The term Montagnard, loosely used, like the word Indian, applies to more than a hundred tribes of primitive mountain people, numbering from 600,000 to a million and spread over all of Indochina. In South Vietnam there are some twenty-nine tribes, all told more than 200,000 people. Even within the same tribe, cultural patterns and linguistic characteristics can vary considerably from village to village. In spite of their dissimilarities, however, the Montagnards have many common features that distinguish them from the Vietnamese who inhabit the lowlands. The Montagnard tribal society is centered on the village and the people depend largely on slash-and-burn agriculture for their livelihood. Montagnards have in common an ingrained hostility toward the Vietnamese and a desire to be independent.

Throughout the course of the French Indochina War, the Viet Minh worked to win the Montagnards to their side. Living in the highlands, these mountain people had been long isolated by both geographic and economic conditions from the developed areas of Vietnam, and they occupied territory of strategic value to an insurgent movement. The French also enlisted and trained Montagnards as soldiers, and many fought on their side.

Since the Rhade (Rah-day) tribe is fairly representative of the Montagnards, a description of the way of life of the villagers will serve as a good example of the environment in which the Special Forces worked in Vietnam. The Rhade were, furthermore,

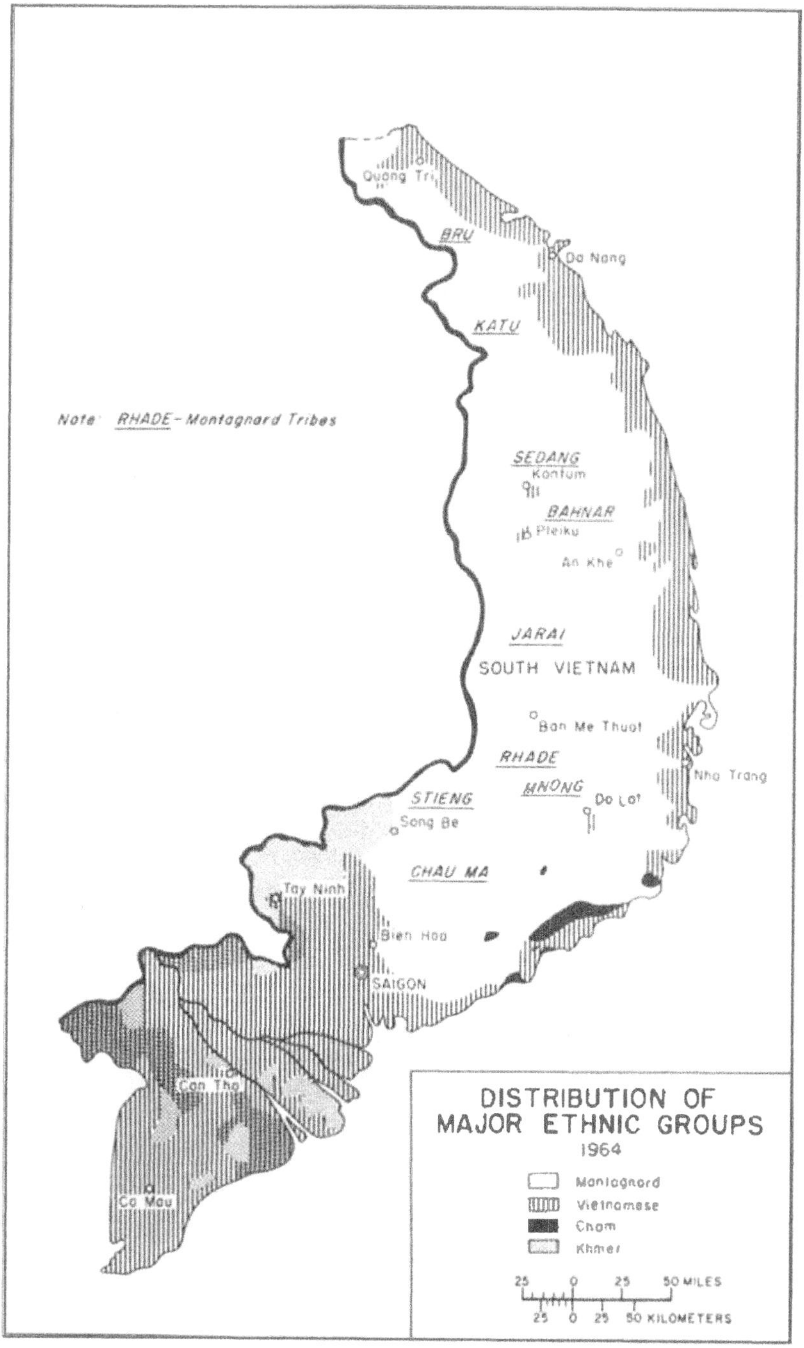
Note: RHADE - Montagnard Tribes
Quang Tri
BRU
Da Nang
KATU
SEDANG
Kontum
BAHNAR
Pleiku
An Khe
JARAI
SOUTH VIETNAM
Ban Me Thuot
RHADE
Nha Trang
STIENG
MNONG
Da Lat
Song Be
CHAU MA
Tay Ninh
Bien Hoa
SAIGON
Can Tho
Ca Mau
DISTRIBUTION OF
MAJOR ETHNIC GROUPS
1964
Montagnard
Vietnamese
Cham
Khmer
25 0 25 50 MILES
25 0 25 50 KILOMETERS

MAP 1

the first to be approached and to participate in the CIDG program. For many years, the Rhade have been considered the most influential and strategically located of the Montagnard tribes in the highlands of Vietnam. (*Map 1*) Mainly centered around the village of Ban Me Thuot in Darlac Province, the Rhade are also found in Quang Duc, Phu Yen, and Khanh Hoa Provinces. While there are no census records for these people, it has been estimated that the tribe numbers between 100,000 and 115,000, with 68,000 living in Ban Me Thuot.

The Rhade have lived on the high plateau for centuries, and their way of life has changed little in that time; whatever changes came were mainly the result of their contact with the "civilized" world through the French. They settle in places where their livelihood can be easily secured, locating their houses and rice fields near rivers and springs. Because they have no written history, not much was known about them until their contact with the French in the early nineteenth century. It is generally agreed that most of their ancestors migrated from greater China, while the remainder came from Tibet and Mongolia.

In order of descending importance, the social units of the Rhade are the family, the household, the kinsmen, and the village. The Rhade have a matrilineal system; the man is the breadwinner, but all property is owned by the wife. The oldest female owns the house and animals. The married man lives with his wife's family and is required to show great respect for his mother-in-law. If a man is rich enough he may have more than one wife, but women may have only one husband. Marriage is proposed by the woman, and the eldest daughter inherits her parents' property.

Building a house is a family enterprise. All members of families who wish to live together pitch in and build a longhouse in accordance with the size of the families. The house is made largely of woven bamboo and is long and narrow, sometimes 400 feet long, with entrances at each end. Both family and guests may use the front entrance, but only the resident families may use the rear. The house is built on posts with the main floor usually about four feet above the ground and is almost always constructed with a north-south orientation, following the axis of the valleys.

The tasks of the man and woman of the family are the traditional ones. The man cuts trees, clears land, weaves bamboo, fishes, hunts, builds houses, carries heavy objects, conducts business, makes coffins, buries the dead, stores rice, makes hand tools and weapons, strikes the ceremonial gongs—an important duty—and is responsible for preparing the rice wine. Authority in the Rhade

RHADE VILLAGE OF BUON ENAO

family is maintained by the man—the father or the grandfather. It is he who makes the decisions, consulting with his wife in most cases, and he who is responsible for seeing that his decisions are carried out. The average Rhade man is between sixty-four and sixty-six inches tall, brown in complexion, and usually broad-shouldered and very sturdy. The men have a great deal of endurance and manual dexterity and have the reputation of being excellent runners.

The woman draws water, collects firewood, cooks the food, cleans the house, mends and washes the clothes, weaves, makes the traditional red, black, yellow, and blue cotton cloth of the Rhade, and cares for the children. The women sit on the porch (the *bhok-gah*) of the longhouse to pound the rice with a long pole and a wooden mortar.

The life of the Rhade is governed by many taboos and customs. Outsiders are expected to honor these, and therefore delicacy was required of Special Forces troops who dealt with the Rhade and other tribes. Healing is the responsibility of the village shaman, or witch doctor, and the general state of health among the Rhade is poor. Religion is animistic—natural objects are thought to be inhabited by spirits—but the tribe also has a god (Ae Die) and a devil (Tang Lie).

The Rhade tend toward a migratory existence. Once they have used up the soil's vitality in one area, they move their village to a new place, seeking virgin soil or land that has not been used for half a century. At the beginning of the rainy season the people plant corn, squash, potatoes, cucumbers, eggplant, and bananas. Once these crops are in the ground, the rice is planted.

The Rhade proved to be enthusiastic participants in the CIDG program in the beginning because the early projects were, they felt, pleasing to the spirits and helpful to their villages. If these two requirements were satisfied (and in many instances they were not later on), the Rhade, and the Montagnards in general, were quite willing to work hard in the CIDG program.

The Montagnards were not, of course, the only minority group involved in the CIDG program; other groups were Cambodians, Nung tribesmen from the highlands of North Vietnam, and ethnic Vietnamese from the Cao Dai and Hoa Hao religious sects.

The Buon Enao Experiment

With the permission of the Vietnamese government, the U.S. Mission in the fall of 1961 approached the Rhade tribal leaders with a proposition that offered them weapons and training if they would declare for the South Vietnamese government and participate in a village self-defense program. All programs that affected the Vietnamese and were advised and supported by the U.S. Mission were supposed to be accomplished in concert with the Vietnamese government. In the case of the Montagnard program, however, it was agreed that the project would at first be carried out separately instead of coming under the command and control of the Vietnamese Army and its advisers, the U.S. Military Assistance Advisory Group. There was no assurance that the experiment with the Rhade would work, especially in the light of the Vietnam government's failure to follow through on other promises to the Montagnards.

The village of Buon Enao, which had a population of approximately 400 Rhade, was visited in late October of 1961 by a representative of the U.S. Embassy and a Special Forces medical sergeant. During two weeks of daily meeting with village leaders to explain and discuss the program, several facts emerged. Because government forces had been unable to protect the villagers many of them supported the Viet Cong through fear. The tribesmen had previously aligned themselves with the government, but its promises of help had failed to materialize. The Rhade opposed the land development program because the resettlement took tracts of tribal lands and because most American and Vietnamese aid went to the Vietnamese villages. Finally, the discontinuance of the medical aid and educational projects by the Vietnamese government on account of the activities of the Viet Cong had created resentment against both the Viet Cong and the government.

The villagers agreed to take certain steps to show their support for the government and their willingness to co-operate. They would build a fence to enclose Buon Enao as a protection and as a visible sign to others that they had chosen to participate in the new program. They would also dig shelters within the village where women and children could take refuge in case of an attack; construct housing for a training center and for a dispensary to handle the promised medical aid; and establish an intelligence system to control movement into the village and provide early warning of attack.

In the second week of December when these tasks had been completed, the Buon Enao villagers, armed with crossbows and spears, publicly pledged that no Viet Cong would enter their village or receive assistance of any kind. At the same time fifty volunteers from a nearby village were brought in and began training as a local security or strike force to protect Buon Enao and the immediate area. With the security of Buon Enao established, permission was obtained from the Darlac Province chief to extend the program to forty other Rhade villages within a radius of ten to fifteen kilometers of Buon Enao. The chiefs and subchiefs of these villages went to Buon Enao for training in village defense. They too were told that they must build fences around their respective villages and declare their willingness to support the government of the Republic of Vietnam.

With the decision to expand the program, half of a Special Forces A detachment (seven members of Detachment A–35 of the 1st Special Forces Group) and ten members of the Vietnamese Special Forces (Rhade and Jarai), with a Vietnamese detachment commander, were introduced to assist in training village defenders and the full-time strike force. The composition of the Vietnamese Special Forces at Buon Enao fluctuated from time to time but was always at least 50 percent Montagnard. A program for the training of village medics and others to work in civil affairs projects intended to replace the discontinued government programs was also initiated.

With the assistance of the U.S. Special Forces and Vietnamese Special Forces troops who had been introduced in December 1961, and a twelve-man U.S. Special Forces A detachment deployed in February 1962, all forty villages in the proposed expansion were incorporated into the program by the middle of April.

Recruits for both village defenders and the local security force were obtained through local village leaders. Before a village could be accepted as a part of the development program, the village chief was required to affirm that everyone in the village would participate in the program and that a sufficient number of people would volunteer for training to provide adequate protection for the village. The program was so popular with the Rhade they they began recruiting among themselves. One of the seven members of Detachment A–35 had this to say about how the Rhade received the program initially: "Within the first week, they [the Rhade] were lining up at the front gate to get into the program. This kicked off the recruiting program, and we didn't have to do much recruiting. The word went pretty fast from village to village." Part of the project's popularity undoubtedly stemmed from the fact that the Montagnards could have their weapons back. In the late 1950s all weapons, including the crossbow, had been denied to them by the government as reprisal for Viet Cong depredations and only bamboo spears were allowed until the second week in December 1961, when the government finally gave permission to train and arm the village defenders and strike forces. The strike force would maintain itself in a camp, while the village defenders would return to their homes after receiving training and arms.

The American and Vietnamese officials were acutely aware of the opportunity for Viet Cong infiltration and developed control measures to be followed by each village before it could be accepted for the Village Self-Defense Program. The village chief had to certify that everyone in the village was loyal to the government and had to reveal any known Viet Cong agents or sympathizers. Recruits vouched for the people nearest them in line when they came for training. These methods exposed five or six Viet Cong agents in each village and these were turned over to the Vietnamese and Rhade leaders for rehabilitation.

Cadres of Rhade trained by the Vietnamese Special Forces were responsible for training both local security (strike) forces and village defenders, with Special Forces troops acting as advisers to the cadres but having no active role as instructors. Villagers were brought into the center and trained in village units with the weapons they were to use, M1 and M3 carbines. Emphasis was placed on marksmanship, patrolling, ambush, counterambush, and swift response to enemy attacks. While members of a village were being trained, their village was occupied and protected by local security troops. Since no official table of organization and equipment existed, these strike force units were developed in accordance with the manpower available and the estimated needs of the area. Their basic element was the squad of eight to fourteen men, capable of acting as a separate patrol.

Activities within the operational area established in co-ordination with the province chief and Vietnam Army units in the vicinity consisted of small local security patrols, ambushes, village defender patrols, local intelligence nets, and an alert system in which local men, women, and children reported suspicious movement in the area. In some cases, U.S. Special Forces troops accompanied strike force patrols, but both Vietnamese and American policy prohibited U.S. units or individual American soldiers from commanding any Vietnamese troops.

All villages were lightly fortified, with evacuation the primary defensive measure and some use of family shelters for women and children. Strike force troops remained on the alert in the base center at Buon Enao to serve as a reaction force, and the villages maintained a mutually supporting defensive system wherein village defenders rushed to each other's assistance. The system was not limited to Rhade villages in the area but included Vietnamese villages as well.

Logistical support was provided directly by the logistical agencies of the U.S. Mission outside Vietnamese and U.S. Army supply channels. U.S. Special Forces served as the vehicle for providing this support at village level, although U.S. participation was indirect in that distribution of weapons and pay of troops was accomplished through local leaders.

In the field of civic assistance, the Village Self-Defense Program provided community development along with military security. Two six-man Montagnard extension service teams were organized to give the villagers training in the use of simple tools, methods of planting, care of crops, and blacksmithing. Village defender and strike force medics conducted clinics, sometimes moving into new villages and thus expanding the project. The civic assistance program received strong popular support from the Rhade.

The establishment of village defense systems in the forty villages surrounding Buon Enao attracted wide attention in other

Rhade settlements, and the program expanded rapidly into the rest of Darlac Province. New centers similar to Buon Enao were established at Buon Ho, Buon Krong, Ea Ana, Lac Tien, and Buon Tah. From these bases the program grew, and by August 1962 the area under development encompassed 200 villages. (*Map 2*) Additional U.S. and Vietnamese Special Forces detachments were introduced. During the height of the expansion, five U.S. Special Forces A detachments, without counterpart Vietnamese detachments in some instances, were participating.

The Buon Enao program was considered a resounding success. Village defenders and strike forces accepted the training and weapons enthusiastically and became strongly motivated to oppose the Viet Cong, against whom they fought well. Largely because of the presence of these forces, the government toward the end of 1962 declared Darlac Province secure. At this time plans were being formulated to turn the program over to the Darlac Province chief and to extend the effort to other tribal groups, principally, the Jarai and the Mnong.

Command and Control During the Buon Enao Period

In the course of the Buon Enao experiment, the command and control structure of the U.S. Special Forces underwent a number of changes. The expansion of the Buon Enao project and the training of Vietnamese Special Forces in 1962 necessarily involved an increase in the number of Special Forces troops needed to do these jobs. This buildup of the U.S. Special Forces generated the need for a Special Forces headquarters in Vietnam, and, with the establishment of the U.S. Military Assistance Command, Vietnam, in February 1962, made co-ordination necessary between the U.S. Mission, which was running the CIDG program and controlling the Special Forces involved in it, and the Military Assistance Command, Vietnam.

In February 1962 there was one full U.S. Special Forces A detachment deployed in Darlac Province on the Buon Enao project. When the Military Assistance Command, Vietnam, was established, with General Paul D. Harkins commanding, a special warfare branch was included in the J–3 staff section. In May a joint agreement between the U.S. Mission and Military Assistance Command, Vietnam, was made to co-ordinate the CIDG program between them. The U.S. Mission initially retained complete responsibility for both the logistical and operational aspects of the program. The counterpart organization to the joint U.S. Mission and Military Assistance Command, Vietnam, structure was the

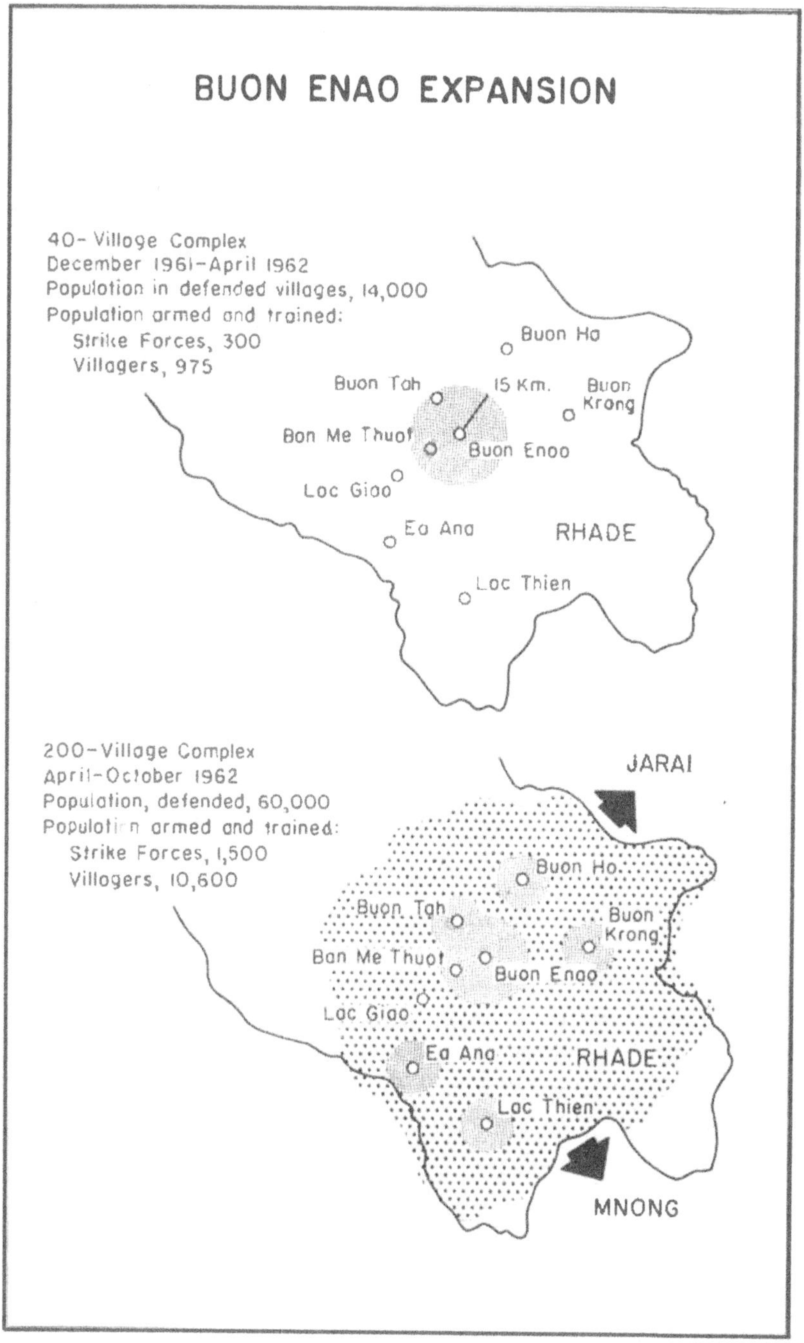
BUON ENAO EXPANSION
40- Village Complex
December 1961-April 1962
Population in defended villages, 14,000
Population armed and trained:
Strike Forces, 300
Villagers, 975
Buon Ho
Buon Tah
15 Km.
Buon Krong
Ban Me Thuot
Buon Enao
Loc Giao
Ea Ana
RHADE
Loc Thien
200-Village Complex
April-October 1962
Population, defended, 60,000
Strike Forces, 1,500
Villagers, 10,600
JARAI
Buon Ho
Buon Tah
Buon Krong
Ban Me Thuot
Buon Enao
Loc Giao
Ea Ana
RHADE
Loc Thien
MNONG

MAP 2

Vietnamese Special Forces under the control of the Vietnam government. In July the U.S. Department of Defense made the decision to transfer complete responsibility for Special Forces operations to the Military Assistance Command, Vietnam, thus making the Army responsible for U.S. support of the Civilian Irregular Defense Group program. The Department of Defense arranged for a colonel qualified in unconventional operations to assume command of the Special Forces in Vietnam and provided for flexibility with respect to supply procedures and the expenditure of funds so that the efficiency and effectiveness of the CIDG program could be maintained. The transfer of responsibility—code-named Operation SWITCHBACK—was to be accomplished in phases and completed by 1 July 1963.

In September 1962 in accordance with Operation SWITCHBACK, Headquarters, U.S. Army Special Forces (Provisional), Vietnam, was activated in Vietnam under the Military Assistance Command. As of October 1962 there were twenty-four U.S. Special Forces detachments in Vietnam. *(Map 3)*

By November of 1962 the U.S. Special Forces organization in Vietnam consisted of one C detachment, three B detachments, and twenty-six A detachments. There was also a headquarters unit in Saigon. The C detachment did not exercise its usual function as an operational control detachment but rather provided augmentation for the headquarters. The normal Special Forces chain of command came into effect.

In the period December 1962 through February 1963, U.S. Army Special Forces (Provisional), Vietnam, assumed full operational control of the Special Forces A detachments in Vietnam. These A detachments had, at this point, established CIDG camps in every one of the four corps tactical zones. A control B detachment was located in each corps tactical zone to co-ordinate with the Vietnamese corps command structure and the senior adviser of the tactical zone and to exercise operational control over subordinate A detachments. Special Forces A detachments were placed on temporary duty in Vietnam from the 1st Special Forces Group on Okinawa and from the 5th and 7th Special Forces Groups at Fort Bragg, North Carolina. By December 1963 Special Forces detachments, working through counterpart Vietnamese Special Forces units, had trained and armed 18,000 men as strike force troops and 43,376 as hamlet militia, the new name for village defenders. Also in February 1963, the U.S. Army Special Forces (Provisional), Vietnam, headquarters was moved from Saigon to Nha Trang. The new location offered two advantages: first,

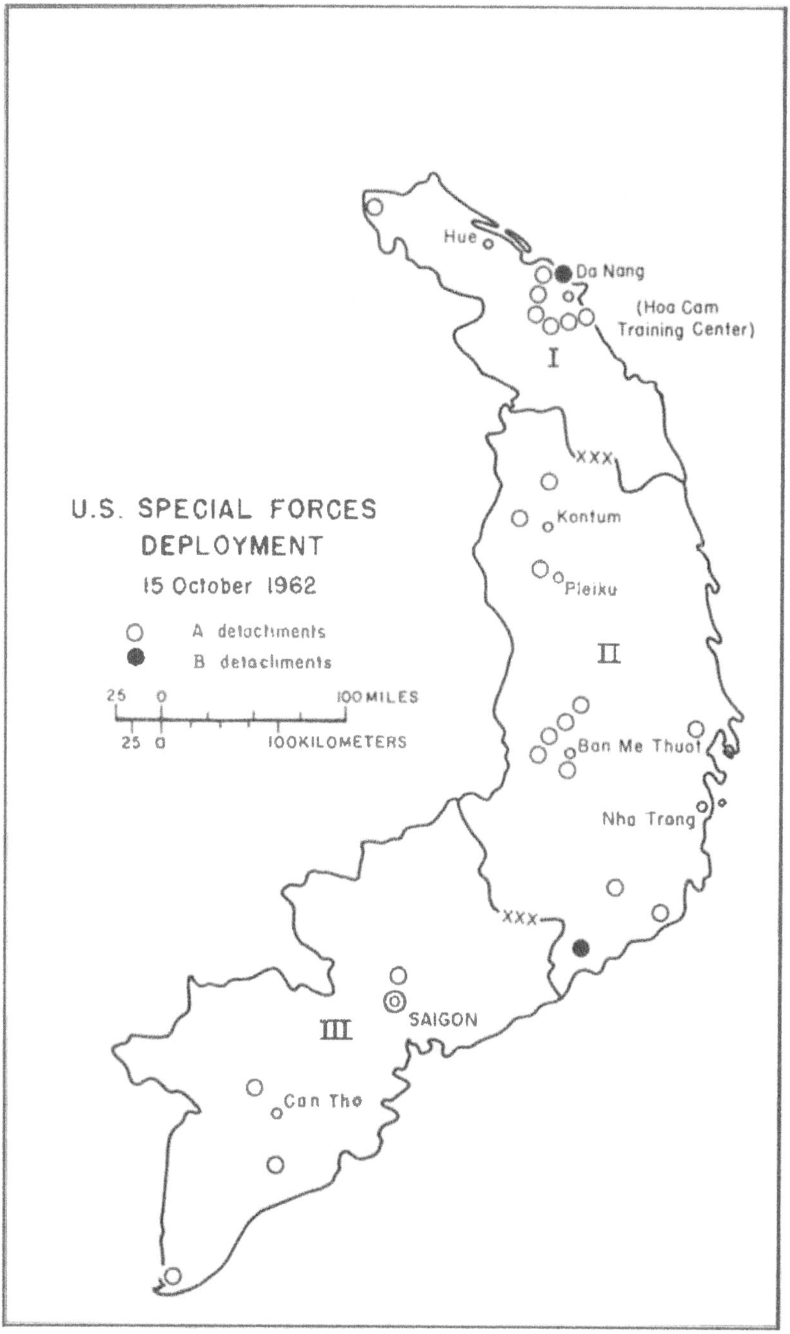

U.S. SPECIAL FORCES
DEPLOYMENT
15 October 1962
A detachments
B detachments
25 0 100 MILES
25 0 100 KILOMETERS
Hue
Da Nang
(Hoa Cam
Training Center)
I
XXX
Kontum
Pleiku
II
Ban Me Thuot
Nha Trang
XXX
SAIGON
III
Can Tho

MAP 3

since it was situated halfway between the 17th parallel and the southern tip of the country, it was more accessible to Special Forces detachments scattered throughout the country, and, second, it afforded good facilities for unloading supply ships from Okinawa across the beach and also had available air, rail, and highway transportation.

In the beginning the program with the Rhade around Buon Enao was not officially called the Civilian Irregular Defense Group program, but was known at different times under various names, one of them area development. The troops were at one time called village defenders, at another time hamlet militia. The term CIDG actually became the official designation for the paramilitary counterinsurgency effort after Buon Enao, when the program began to expand.

The success of the Buon Enao experiment prompted a rapid and wide-ranging expansion of the Civilian Irregular Defense Group program and an accompanying expansion of the Special Forces' role. The expansion period ran from August 1962, when the Buon Enao project was flourishing with a 200-village complex, to the spring of 1965, when substantial numbers of conventional U.S. combat units began to reach Vietnam.

Early Paramilitary Programs

During the period of the Buon Enao experiment with the Rhade, a number of other programs, often independent and unrelated, were initiated by the U.S. Mission in Saigon in an effort to extend government control into areas either lost to the government or under marginal control. All these programs, along with the area development program that extended out of Buon Enao in the spring of 1962 further into the Montagnard region and elsewhere, came to be designated officially as the Civilian Irregular Defense Group program. In these other programs the Special Forces, under the control of the U.S. Mission, conducted paramilitary training programs for the minority groups involved.

In December 1961 the second half of Detachment A-35 (the other half was at Buon Enao) arrived at the Hoa Cam Training Center in Da Nang, where it inaugurated a basic training program together with several specialized programs. Among the paramilitary units trained at Hoa Cam were the mountain commandos, later called mountain scouts. These men were used on long-range missions in remote jungle and mountain areas in order to provide a government presence in the areas and to gather intelligence for the military and civil authorities in their districts. An-

SPECIAL FORCES SOLDIER REMOVES BAZOOKA SHELL *after gunner has jammed it in the tube.*

other program in which the Special Forces functioned as training cadre was the trailwatchers program. The mission of the trail-watchers, later called border surveillance units, was to identify and report Viet Cong movements near the border in their area and to capture or destroy small Viet Cong units when possible. Montagnards living near the border of Vietnam with Cambodia and Laos who participated in the trailwatchers program were trained for eight weeks at Da Nang or at other area development centers. The trailwatchers program is significant in that it produced the border surveillance program, in which the concepts of area development and border surveillance were combined to form one of the most important facets of the CIDG program.

The Special Forces also helped train paramilitary forces in the "fighting fathers" program, wherein resistance to insurgent activity centered on Catholic parish priests and a number of priests under the program made the arming and training of their parishioners possible. The goal, again, was to secure an area for the government of South Vietnam. The Civilian Irregular Defense Group program emerged as an amalgamation of many little programs, all of which aimed at the protection of and development of minority groups against insurgency.

By the end of 1964 the Montagnard program was no longer an area development project in the original sense of the term.

There was a shift in emphasis from expanding village defense systems to the primary use of area development camps or centers (CIDG camps) as bases for offensive strike force operations. At the time the principal task as seen by higher headquarters was to supplement the current government pacification program with intensified counterguerrilla warfare. Security and camp defense took precedence over civic action, and stress was laid on the role of CIDG strike forces as "VC hunters." A second major shift in mission that gave greater importance to border surveillance occurred in 1963 as area development projects were expanded toward the western borders of Vietnam and new CIDG camps were established in border areas. Although area development continued in other localities and was combined with border surveillance when feasible, border surveillance received greater emphasis in 1964.

With the expansion of the CIDG effort among the Montagnards, other tribal groups were drawn in and more Special Forces detachments became involved. New projects were not concentrated in a specific area but were dispersed and scattered throughout the country. The original Buon Enao complex expanded in Darlac Province. Projects were also initiated to recruit tribes in I Corps Tactical Zone and the northern regions of II Corps Tactical Zone (in Kontum and Pleiku Provinces). Support was given to the Catholic youth program in the Mekong Delta. All this expansion involved area assessment and the setting up of area development centers. (*See Map 3.*)

A CIDG area development center consisted essentially of a secure base of operations at which village defenders and strike forces recruited from nearby villages were trained. As at Buon Enao, the village defenders (later known as hamlet militia) volunteered to come to the base camp for training and to receive weapons while their villages were protected by the strike force. Supervised and assisted by strike forces, village defenders, after returning to their homes, were expected to patrol in defense of their villages and their immediate area. They received no pay except during their period of training. Each village development area was protected by paid, full-time strike force troops. Operating in units of platoon or company size, they conducted aggressive patrolling throughout the operational area, assisted villages under attack, set up ambushes, and checked village defense procedures. In most cases, strike force troops were paid in accordance with Vietnam Army pay scales, although in the early stages of the program higher pay was sometimes used as an incentive for enlistment.

The general mission of an area development center or CIDG camp was to train strike forces and village defenders; bring the local populace under the influence of the South Vietnam government; employ paramilitary forces in combat operations to reinforce organized hamlets, carry out interdiction activities, and conduct joint operations with Vietnamese Army units when such operations furthered the CIDG effort; conduct psychological operations to develop popular support for the government; establish an area intelligence system including, but not limited to, reconnaissance patrols, observation posts, and agent informant networks; conduct a civic action program; and, where appropriate, establish a border screen in sectors along the Republic of Vietnam international border. During the development phase, all reasonable means were to be taken to improve the economic status of the local population by purchasing local materials and hiring local labor for the construction and operation of the camp. At this time, the CIDG area development camp plans called for eventual turnover and integration into the national strategic hamlet program after the area was "pacified." The overthrow of the Diem regime in 1963, however, altered those plans.

Throughout the period 1961–1963, the U.S. Special Forces units were conducting training programs, both in support of the Military Assistance Advisory Group—for example, Ranger training—and in support of the paramilitary programs. The major mission, however, of the U.S. Special Forces after Buon Enao was to establish base camps and conduct operations in support of the area development program under the U.S. Mission. At the camps the U.S. Special Forces advised their counterpart Vietnamese Special Forces detachments, provided operational assistance when required, and served as a channel for the logistical and financial support provided by the U.S. Mission.

Operation Switchback: November 1962–July 1963

In accordance with Operation SWITCHBACK, the Army began assuming responsibility for U.S. participation in the Civilian Irregular Defense Group program in November 1962 by first taking over training and operations. By 30 June 1963 the Military Assistance Command, Vietnam, was fully responsible for logistics and funding and in July the administration of the Civilian Irregular Defense Group program belonged to the command completely. There were, however, some paramilitary projects that had not yet been officially incorporated into the CIDG program and therefore did not come under Military Assistance Command con-

HELICOPTER LIFTS OFF AFTER DROPPING SUPPLIES *to isolated Special Forces team.*

trol. The most significant of these was the border surveillance program, which was not incorporated into the CIDG program until October 1963. At that time responsibility passed from the U.S. Mission to the U.S. Military Assistance Command, Vietnam. Border surveillance sites were then considered as CIDG projects with a subsidiary mission of border surveillance and were administered in accordance with the CIDG program.

The unusual logistic support system used by the Special Forces in the conduct of the CIDG program had a great deal to do with its successes. The U.S. Army Support Group, Vietnam, provided the normal supply support for U.S. Special Forces detachments. Through interagency agreements, the Army, incorporating many features of the U.S. Mission's logistical system, gave the U.S. Special Forces a direct overseas procurement capability, authorized local purchases from current operating funds at all U.S. Special Forces levels, allowed for informal justification for unusual items or quantities, dropped formal accountability for items on shipment to Vietnam, and devised what came to be known as "quick-reacting supply and procurement procedures." The Counterinsurgency Support Office was established in the G–4 section of Headquarters, U.S. Army, Ryukyu Islands, on Okinawa, to control and expedite procurement and shipment of material supplied from outside Vietnam. During Operation SWITCHBACK, a monthly average of approximately 740 tons of equipment and supplies was airlifted from Saigon and Da Nang to Special Forces A detachments.

By the end of 1962, only one year after the initiation of the Buon Enao project, 6,000 strike force troops and 19,000 village defenders and hamlet militia had been trained. Other irregulars trained included 300 border surveillance troops, 2,700 mountain scouts, and approximately 5,300 Popular Forces troops. (*Table 1*)

The expansion of the CIDG program from 1 November 1962 to 1 July 1963, the end of Operation SWITCHBACK, was fairly rapid. Approximately forty CIDG camps were opened and eight closed. The rapidity of this expansion did not permit the kind of development that took place at Buon Enao, where a great deal of time was taken to prepare the area and the people for military activity in the CIDG program. This time the emphasis was on speed. The usual approach was to establish security first, undertake civic action later, and work through province and district chiefs instead of tribal leaders. In general these projects were not as successful as the Buon Enao experiment. In many areas Viet Cong control was stronger than at Buon Enao, making recruitment difficult. Often the tribal groups were not as advanced as the Rhade. Strike forces frequently had to be moved from their home areas in order to establish a new camp. It was also during this period that emphasis shifted from the establishment of mutually supporting village defense systems to carrying out offensive strike force operations in order to open up and then secure an area. Finally, increased emphasis was placed on selecting area development campsites near the borders so that strike forces could be assigned a border surveillance mission in addition to the task of clearing and securing the assigned operational area. (*Map 4*)

The primary U.S. Special Forces mission during Operation SWITCHBACK continued to be the training of strike force troops and hamlet militia. By June 1963 approximately 11,000 strike force and 40,000 hamlet militia from over 800 villages had undergone training that averaged about six weeks for strike force troops and two weeks for hamlet militia. The Special Forces also continued to conduct training for the border surveillance and mountain scout programs.

TABLE 1—DISPOSITION OF U.S. SPECIAL FORCES DETACHMENTS, DECEMBER 1962

Province	CIDG Campsite	Mission
	I Corps	
Quang Tri	Khe Sanh	BS,[a] CIDG
Quang Nam	Da Nang	Control (B det)
	Hoa Cam	CIDG, MS,[b] AR,[c] misc.
	Phu Hoa	CIDG
Quang Tin	Tra My	CIDG
	II Corps	
Quang Ngai	Tra Bong	CIDG
	Ba To	CIDG
	Mang Buk	CIDG
Kontum	Dak Pek	CIDG, BS
	Tanh Canh	CIDG, BS
Pleiku	Pleiku	Control (B det)
	Plei Mrong	CIDG, BS
	Plei Yt	MS (1 det)[d]
Binh Dinh	Van Canh	CIDG
Phu Bon	Cheo Reo	CIDG
	III Corps	
Darlac	Ban Don	BS, CIDG
	Buon Enao	CIDG
	Buon Dan Bak	CIDG
	Krong Kno Valley	CIDG
Tuyen Duc	Serignac Valley	CIDG
Khanh Hoa	Nha Trang	Control (Alt Hq) } B det
		Control (III Corps) } B det
Ninh Thuan	Phuoc Thien	CIDG
Binh Thuan	Song Mao	CIDG, SF,[e] misc
	Capital Region	
Saigon	Saigon	Command (Hq det)
Cia Dinh	Thu Duc	AR, misc
	IV Corps	
Phong Dinh	Can Tho	Control (B det)
Ba Xuyen	Du Tho	CIDG
An Giang	Chau Long	CIDG (1 det)[d]

[a] Border surveillance.
[b] Mountain scout.
[c] Airborne Ranger, Vietnam Army.
[d] On-the-job training.
[e] 77th Special Forces, Vietnam Army.

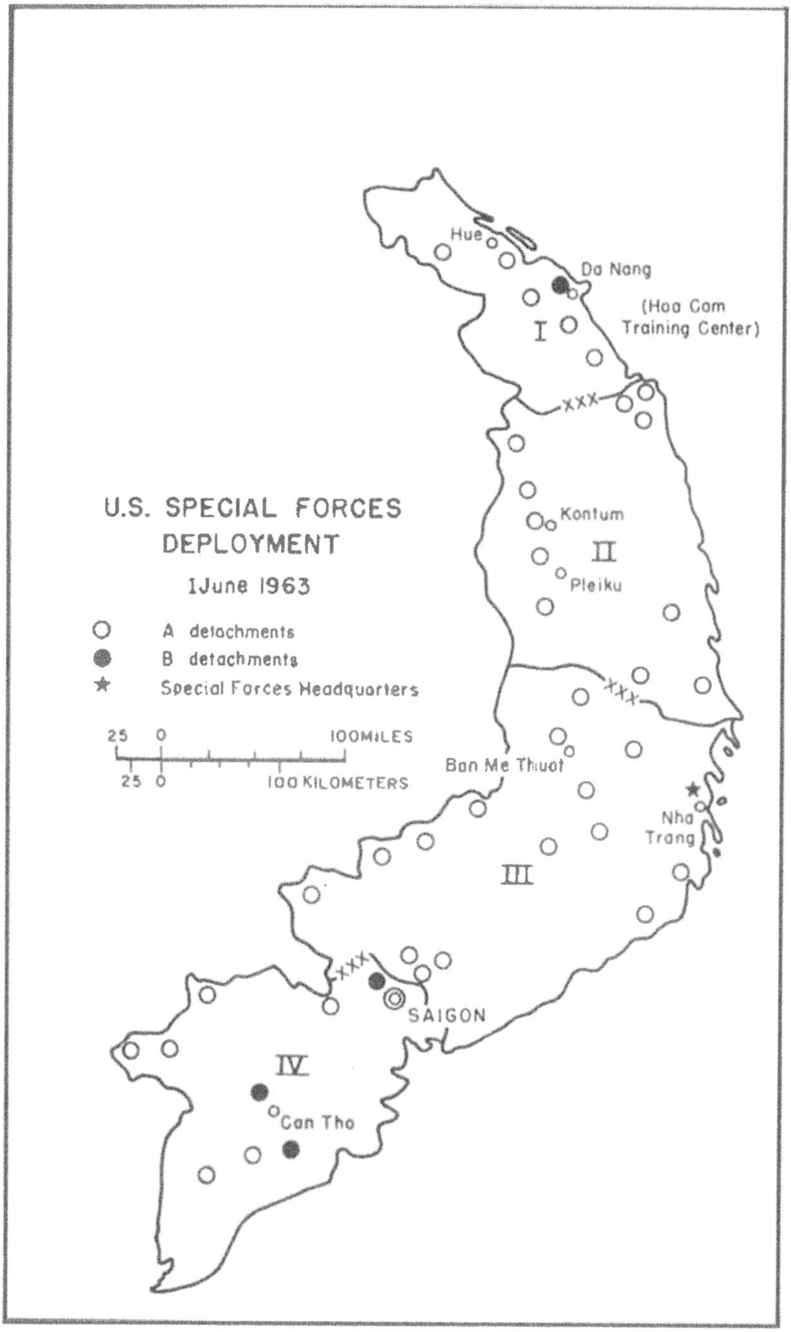
U.S. SPECIAL FORCES
DEPLOYMENT
1June 1963
A detachments
B detachments
Special Forces Headquarters
25 0 100MILES
25 0 100 KILOMETERS
Hue
Da Nang
(Hoa Cam
Training Center)
I
II
III
IV
Kontum
Pleiku
Ban Me Thuot
Nha
Trang
SAIGON
Can Tho

MAP 4

Offensive actions conducted by CIDG strike forces during Operation SWITCHBACK included ambushes, reconnaissance patrols, and combat patrols within each camp's operational area. By June 1963 the CIDG camps in II Corps had completed the training of enough strike force troops to enable U.S. Special Forces (Provisional), Vietnam, to shift emphasis from training to operations against the Viet Cong. There were joint operations with Vietnam Army units and also combined operations using CIDG forces from different camps. In June, for example, four companies of strike forces from the camps at Dak To, Plei Mrong, and Polei Krong were combined for an operation. The role U.S. Special Forces men played in many of these operations was much more positive in terms of direction and control than their role as advisers would indicate; in many instances Vietnamese Special Forces troops were unable to carry out an operation and U.S. Special Forces men were obliged to take over command. Vietnamese Special Forces officers and noncommissioned officers were not as well trained as their U.S. counterparts, and were, furthermore, often unwilling to carry out offensive operations with the civilian irregulars. This is not to say that they were afraid. Most had seen a great deal of fighting. They were just not interested in, or even remotely enthusiastic about, the CIDG program. From the point of view of the Vietnam Special Forces and the government the CIDG program was an American project. The failure of the turnover and conversion of camps grew largely out of this unenthusiastic attitude on the part of the Vietnamese. In any case, combat operations often placed an exceptional burden on U.S. Special Forces soldiers. A typical operation might involve a company of indigenous CIDG Montagnard strike force troops and one or two Special Forces men—an officer and a noncommissioned officer. If the Vietnamese Special Forces troops were along, then the Green Beret was an adviser. But in a firefight he often became the commander, and the men he had around him were not Americans. The Special Forces has more than a few decorations for acts of individual heroism under such circumstances. (*See Appendix B.*)

Viet Cong reaction to the expansion of the CIDG program in many instances took the form of mere harassment or occasional probing fire. During Operation SWITCHBACK, however, Viet Cong opposition increased. Every CIDG camp experienced some sort of enemy fire. The one instance of an attack in strength occurred on 3 January 1963 when two reinforced Viet Cong companies, with the assistance of at least thirty-three penetration agents in the strike force, attacked and overran the camp at Plei Mrong. The cumulative effect of these attacks was an increase in camp security; specifically, Military Assistance Command, Vietnam, in January 1963 issued a directive to all A detachments that called for a secondary defensive system inside the outer perimeter, along with other measures.

Finally, it was during Operation SWITCHBACK that the Special Forces began to receive augmentation. Two U.S. Navy Construction Battalion (Seabee) technical assistance teams arrived in January on a six-month temporary duty basis and were employed extensively in airfield construction. Two U.S. Army five-man Engineer Control and Advisory Detachments also came in to assist the Special Forces in constructing roads, schools, drainage systems, and other civic action projects. Augmentation also included direct assistance in civil affairs and psychological operations. In June 1963, men specially trained in civil affairs and psychological operations were assigned to Special Forces (Provisional), Vietnam, and two-man teams were dispatched to Special Forces B detachments.

The Turnover of Buon Enao

The problems encountered in turning over the Buon Enao project to the Vietnamese proved to be the same problems which arose every time turnover was attempted in the CIDG program. A discussion of the Buon Enao experience, therefore, is illustrative of the broader turnover experience.

The so-called turnover of a CIDG camp consisted of nothing more than "turning over complete authority and responsibility for the camp" to the Vietnamese Special Forces. The CIDG forces present in the camp maintained their CIDG status. The opposite was true in a so-called conversion: civilian irregulars were converted into regular Vietnamese soldiers, Regional or Popular Forces or Vietnamese Army, and lost their CIDG status. Conversion was not very popular among the CIDG, although eventually most men did convert.

The concept of the CIDG program provided that when an area could be considered secure or became accessible to Vietnam Army units and government agencies, it would be turned over to provincial control. By the end of Operation SWITCHBACK, July 1963, the expanded CIDG program was still in the developmental stage—only the villages in the Buon Enao complex were considered secure and ready for turnover. Nevertheless, the turnover of Buon Enao was a failure. A 5th Special Forces Group synopsis of the CIDG program concluded: "By the end of 1963, the Buon

Enao complex was disorganized and most of its effectiveness had been lost."

The turnover at Buon Enao gave the first indication of two major problems that would arise later in the transfer of men trained by Special Forces to the government of Vietnam. The first of these was the lack of preparation on the part of the government for taking over and continuing area development projects. The second was the reluctance of strike force troops to be integrated into conventional Vietnamese units. During Operation SWITCHBACK some camps were closed out before the assigned mission was accomplished because of the lack of CIDG potential in the area, change of mission, or greater need elsewhere for the strike force personnel.

In September 1962 the Darlac Province chief agreed to accept thirty-two of the 214 villages in the Buon Enao complex. These villages were considered secure. It was planned to turn over an additional 107 villages at the end of March 1963 and the remaining villages by the end of June 1963.

The government arranged a very rigid schedule for the Buon Enao turnover. The province chief and the Vietnamese Special Forces were given orders to carry out the schedule to the letter. The U.S. Special Forces teams on the sites were therefore unable to alter the schedule, although there is evidence of their considerable apprehension over the possible consequences of an unprepared turnover—apprehension that apparently was not shared by the government. Despite the formal co-ordination that took place, actual on-the-scene planning and execution of turnover was handled unilaterally by the Vietnamese. The commanding officer of U.S. Special Forces (Provisional), Vietnam, was not shown the turnover plan in advance.

By the end of 1962 the chief of Darlac Province had accepted the thirty-two villages, but since he was unable to support them financially or logistically the villages were turned over on paper only. The Special Forces had to continue to support the villages and pay the strike force and other costs. On 20 March 1963 the second lot of 107 villages was turned over with province support of the original thirty-two still not forthcoming. The last 139 villages were to be assimilated into the strategic hamlet program, but because these hamlets had not yet been approved for support by regular U.S. Vietnamese funds, U.S. Mission funds continued to support all 214 villages in the complex for the next few months.

In April 1963, 604 of the 900-man Buon Enao Strike Force were turned over to the province chief to be used for the CIDG program and border surveillance, and one company was sent to open a new camp at Bu Prang in Quang Duc Province. The day after the turnover, the province chief moved the 604 CIDG troops from Buon Enao to Ban Me Thuot for indoctrination, thus leaving the complex without a strike force during the hours of darkness. These actions were taken unilaterally by the Vietnamese and apparently without any indoctrination or psychological preparation of the strike force. Although the province chief had assumed responsibility for the pay of strike force troops on 30 April 1963, they were still unpaid by 26 July and seemed ready to desert, despite the prospect of higher pay scales at the Vietnam Army rate. Neither had the village health workers trained at Buon Enao been paid. A serious situation was narrowly averted when the Special Forces provided back pay to strike force troops and village health workers from CIDG funds.

Once the strike force troopers left Buon Enao for indoctrination, they did not return but were transferred to other parts of the province, and in the process their unit integrity was destroyed. The dependents of the strike force also started to leave; and there were other disturbing developments that added to the bewilderment and discontent of the Rhade, who had come to view Buon Enao as the source and symbol of the entire program. For example, the dispensary facilities at Buon Enao, which had played a major role in the initiation of the project, were dismantled and moved to Lao Tien and Buon Ho.

Concern in Saigon about the large number of weapons distributed to the Rhade resulted, in December 1962, in a government order to reduce the number of weapons by 4,000. Difficulty was encountered in collecting the weapons because the tribesmen had received no instructions to turn them in. The order to do so appeared to them to be inconsistent with what they had been told, namely, that the weapons given them were for the defense of their villages and families. At the time of the turnover, there were still 2,000 more weapons in the province than Saigon regulations permitted, and there were further collections. The Special Forces did not participate in the collection of weapons. Disillusionment following the turnover of Buon Enao may have contributed to the Montagnard uprising which took place in late September 1964.

The reasons for the failure of the Buon Enao turnover can be summarized as follows: mutual suspicion and hostility between the Rhade and Vietnamese province and district officials; overly generous distribution by U.S. agencies of weapons and ammunition to tribesmen whose reaction to government enforced repossession of some of the weapons was understandably hostile; apparent disregard on the part of the Vietnam government for the interests, desires, and sensitivities of the Montagnards; inadequate Vietnamese government administrative and logistical support; and, finally, failure of U.S. authorities to anticipate these difficulties and avoid them.

Although the transfer of considerable assets took place during the early years, there were only rare instances of turnover where the mission of pacification had been accomplished and where troops were trained and ready to carry on without U.S. supervision. Of the eighty-two CIDG camps that were established prior to October 1964, more than forty were closed out or turned over by that date.

The higher priority given the border surveillance mission after July 1963 caused a shift of the principal CIDG effort from the interior to border sites and was the reason for the turnover or closeout of seventeen of the eighteen camps relinquished between August 1963 and March 1964. Inadequate initial area assessment led in some cases to the selection of unproductive sites that later had to be relocated. Some camps were situated on indefensible terrain or had limited CIDG potential. Other camps were moved or closed out altogether because of the discontent of the strike force, which had been recruited from a distant area because of the lack of local resources.

CHAPTER III

The CIDG Program Under the U.S. Military Assistance Command, Vietnam: July 1963–May 1965

From 1963 to 1965 the Civilian Irregular Defense Group program continued to develop rapidly and there were changes not only in its organization and command structure but also in its mission. New combat techniques evolved and the integral role the CIDG would play in the U.S. counterinsurgency effort in Vietnam began to take shape. The Montagnard uprising forced a re-evaluation of the program and placed new emphasis on developing satisfactory turnover procedures. The U.S. Special Forces in Vietnam became the 5th Special Forces Group (Airborne). Attention continued to be devoted to civic action and psychological operations with further augmentation of the Special Forces in these areas, and there were new developments in the logistical support system for the CIDG program.

With the phase-out on 1 July 1963 of the U.S. Mission's logistical responsibility for the CIDG program, U.S. Military Assistance Command, Vietnam, assumed complete responsibility for all Special Forces activities in Vietnam, with the exception of four A detachments engaged in surveillance along the Laotian border. These detachments passed to the control of the Military Assistance Command in November 1963 when it became responsible for everything, including the entire border surveillance mission, in the CIDG program. In order to achieve closer co-ordination between Special Forces (Provisional), Vietnam, and Military Assistance Command, Vietnam, a tactical operations center, composed of the S–2 and S–3 sections from Nha Trang, was established in Saigon in March 1964. The logistical support center S–1 and S–4 staff sections and the headquarters section remained in Nha Trang. This split command remained unchanged until the end of October 1964.

Until May 1964 the Special Forces chain of command—from headquarters at Nha Trang to B detachments at the four corps headquarters to operational A detachments---remained unchanged and distinct within the Military Assistance Command, Vietnam, chain. With the growth of the U.S. advisory effort and the expansion of counterinsurgency programs, there was an increasing need for better co-ordination at all levels, particularly in the field where in many areas CIDG camps were near Vietnamese Army and provincial units. Accordingly, on 1 May 1964, in order to integrate more effectively the CIDG program within the countrywide pacification program, operational control of Special Forces A and B detachments was transferred to the Military Assistance Command senior advisers in each corps tactical zone. The Vietnamese Special Forces were already under corps command, so both the U.S. Special Forces and the Vietnamese Special Forces were responsive to the same command levels.

During Operation SWITCHBACK, a number of CIDG camps had been placed along the Laotian-Cambodian border. *(See Map 4.)* Emphasis was placed on expanding the Montagnard area development camps toward the border to provide border-screening forces. After Operation SWITCHBACK, U.S. Military Assistance Command, Vietnam, placed even greater emphasis on border surveillance and control. The Border Surveillance–Control Operating Concept drafted by Special Forces (Provisional), Vietnam, specified the following missions for U.S. and Vietnamese Special Forces: recruit and train personnel to serve in border surveillance and control units in populated areas; establish intelligence nets in the border areas to detect infiltration; direct psychological indoctrination and civic action programs in the border control zone: gain control of the international border little by little and gradually expand small secure areas until the border zone should be permanently under the control of the Border Command; and conduct guerrilla warfare—long-range patrol activities to deny the border areas to the Viet Cong by detection, interdiction, harassment, and elimination of the infiltration routes parallel to or through the border control zone.

In November 1963 the U.S. Mission's responsibility for border surveillance was terminated and the mission assigned to Special Forces (Provisional), Vietnam. From this date, priority was given to the establishment of CIDG camps near the border where they could carry out a border surveillance or area development center and border surveillance mission, depending on the density of the population in the area. CIDG and Special Forces did not, however, have the entire mission. The Vietnamese Pacification Plan called for use of both CIDG camps and Vietnamese Ranger battalions.

SPECIAL FORCES CAMP *with team house on the left and mortar pits in flood dikes on the right. Guy wires support antenna pole in center.*

Other government forces manned approximately seventy-six border outposts, mainly of a static, defensive type, with a total strength of 3,860 men. In contrast, the CIDG concept involved active patrolling by screening forces, which often operated from forward operating bases when the CIDG campsite was some distance from the border. By 1 July 1964, twenty-five border projects employing eighteen Special Forces detachments and 11,250 strike force troops had been initiated.

Planning and organization for border surveillance was based on the planning factor that a CIDG camp could exert appreciable influence over the area within a ten-kilometer radius of the camp. The ideal distance, then, between border surveillance camps would presumably have been twenty kilometers. The addition of two border camps in IV Corps in late 1964, which resulted in an average distance of twenty-seven kilometers between camps, was the closest Special Forces (Provisional), Vietnam, was able to come to this spacing interval.

A strike force ceiling of 20,000 was set in November 1963 to be attained by July 1964. In March 1964, a table of organization and equipment for a CIDG light guerrilla company was issued. (*Chart*

CHART 5—ORGANIZATION OF CIVILIAN IRREGULAR DEFENSE GROUP COMPANY IN 1964

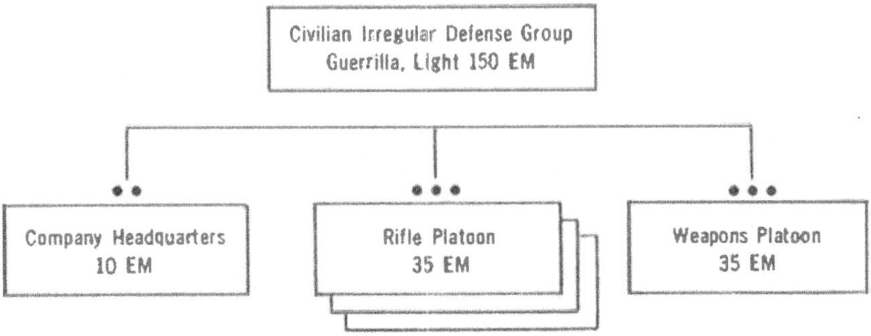

Major Equipment Item	*No.*
Carbine, .30-caliber, M1	103
Launcher, grenade, carbine	24
Machine gun, light, .30-caliber	3
Mortar, 60-mm.	3
Pistol, .45-caliber	1
Radio, HT–1	20
Radio, TR–20	2
Rifle, automatic, Browning, .30-caliber	18
Submachine gun, .45-caliber, M3	29

5) The conventionalization of CIDG forces had begun. Before this, the Special Forces troops and their counterparts had enjoyed considerable leeway in determining the strength of the CIDG forces at the various camps. The number of CIDG troops at any camp reflected, to the extent possible, the nature of anticipated operations as well as the possible recruits in the area.

The importance of the area development mission continued to fall off during this period with the shift of the program to the borders. The area development mission was still in force in border surveillance posts where there were adequate population resources, but many camps were located in such isolated regions as I Corps in the north where no such resources existed. In cases such as these, irregulars often had to be hired for pay and transported, along with their families, to the campsites. Very few hamlet militia were trained; the emphasis would be on off-site offensive operations, with a corresponding increase in advanced individual training and tactical practice for the civilian irregulars involved.

Another trend toward the conventional employment of CIDG forces was their increased use in joint operations with Vietnam Army and other government units after assumption of their operational control by corps. In such operations, CIDG forces were used as regular troops in activities for which they had not been intended and, in many cases, for which they had not been trained or equipped.

The ethnic composition of the Civilian Irregular Defense Group also expanded during this period to embrace Cambodians and the Cao Dai and Hoa Hao, religious sects in the Mekong Delta. These groups, like the Montagnards, did not always have the same goals as the government, which accepted their inclusion in the program reluctantly. (*Table 2*)

The initiation of new projects and the construction of camps constituted the major portion of the Special Forces effort during this period, largely as a result of the shift of attention to the borders under the border surveillance mission. Approximately thirty camps were opened. (*Map 5*) At that time, there were forty-four A detachments in Vietnam with four B detachments controlling them—one in each corps tactical zone. Another B detachment was assigned to the Vietnamese Special Forces Training Center at Dong Ba Thin, just south of Nha Trang. A corollary to the opening of new camps was the requirement to close a number of other sites and a subsequent relocation of these troops to camps nearer the border. The Special Forces participated in the turnover or closeout of approximately thirty CIDG camps during this period. In areas that were considered secure, the paramilitary forces were turned over to provincial officials; in others, notably I Corps, the number of Montagnards available for CIDG recruitment was always very small and rapidly absorbed.

The assumption of the border surveillance mission and the de-emphasis on area development resulted in a training requirement for strike force troops only. Very few hamlet militia were trained after November 1963, almost none after April 1964. Basic training continued for strike force recruits, and much effort was devoted to retraining former members of the CIDG forces.

Combat Operations

There were a number of developments in the area of combat operations during the July 1963–May 1965 period, just before the extensive buildup of conventional U.S. forces. These were significant in that they laid the foundation for the combat operations that would characterize the Special Forces civilian irregulars in the years 1965–1970.

Emphasis continued to be placed on the offensive role of strike

TABLE 2—ETHNIC BACKGROUND OF STRIKE FORCES AT U.S. SPECIAL FORCES CIDG CAMPS, OCTOBER 1964

Site	Composition
I Corps Tactical Zone	
Da Nang	Nung
Ta Ko	Vietnamese
Kham Duc	Vietnamese
Khe Sanh	Vietnamese, Bru
A Shau	Vietnamese, Tau-Oi
A Ro	Vietnamese
II Corps Tactical Zone	
Gia Vuc	Hre
Pleiku	Nung, Rhade, Vietnamese
Dong Ba Thin	Vietnamese, Cham, Tuong, Mien
Plei Do Lim	Jarai, Bahnar, Nung
Buon Mi Ga	Mnong
Ban Don	Rhade, Mnong, Jarai, Thai
Bu Prang	Rhade, Mnong
Dak Pek	Sedang, Jarai, Halang, Vietnamese, Nung
Nha Trang	Vietnamese, Cham, Rhade, Raglai, Nung
Polei Krong	Rongao, Jarai, Bahnar
Plei Djereng	Jarai, Bahnar
Duc Co (Chu Dron)	Jarai, Nung
Kannack	Bahnar, Rhade, Jarai, Bong, Mien, Nung
Dong Tre	Bahnar, Nung
Buon Brieng	Rhade
Plei Me	Jarai, Nung
Plei Mrong	Jarai, Bahnar, Rhade
Dak To	Sedang, Nongao, Bahnar
Plei Ta Nangle	Bahnar, Jarai, Nung
Phey Srunh	Koho, Ma, Chil
Buon Beng (Cheo Reo)	Jarai, Bahnar, Drung
III Corps Tactical Zone	
Bu Chia Map	Vietnamese, Cambodian, Stieng
Suoi Da	Vietnamese, Cambodian
Ben Cat	Vietnamese, Cambodian, Nung
Bu Dop	Stieng
Minh Thanh	Vietnamese, Cambodian
Trang Sup	Vietnamese, Cambodian
Nui Ba Den	Vietnamese, Cambodian
Loc Ninh	Vietnamese, Cambodian
Long Thanh	na
Phuoc Vinh	Vietnamese, Nung, Cambodian
IV Corps Tactical Zone	
Moc Hoa	Vietnamese, Cambodian, Nung
An Long	Vietnamese, Cambodian, Hoa Hao
Tinh Bien	Vietnamese, Cambodian
An Phu	Hoa Hao
Don Phuoc	Vietnamese, Hoa Hao, Nung
To Chau	Vietnamese, Cambodian
Vinh Gia	Vietnamese, Cambodian

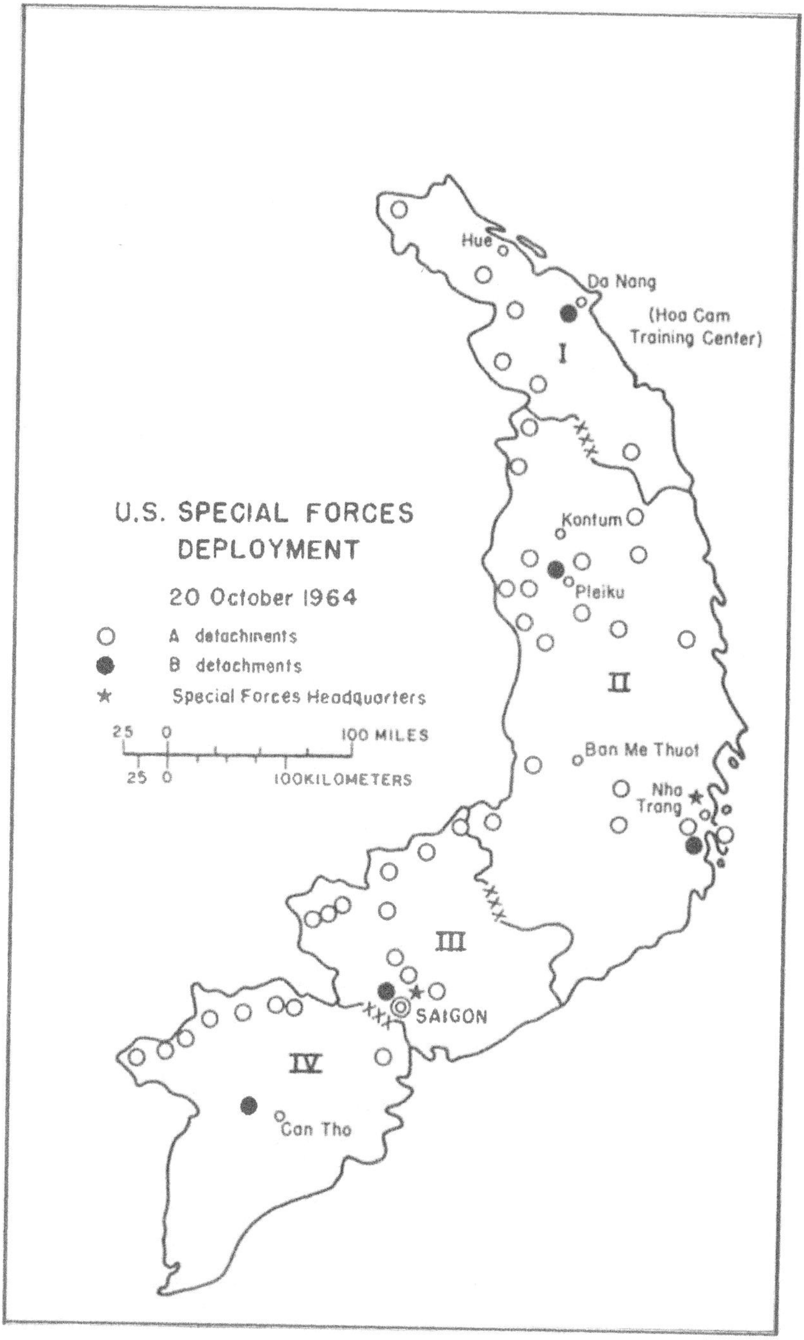
U.S. SPECIAL FORCES
DEPLOYMENT
20 October 1964
A detachments
B detachments
Special Forces Headquarters
25 0 100 MILES
25 0 100KILOMETERS
Hue
Da Nang
(Hoa Cam
Training Center)
I
Kontum
Pleiku
II
Ban Me Thuot
Nha
Trang
III
SAIGON
IV
Can Tho

MAP 5

forces in both area development and border surveillance operations. The border surveillance operations were not as effective as the planners had hoped. The camps were too far apart, averaging twenty-eight miles between each, and the indigenous CIDG platoon and squad leaders were not really capable of conducting their own independent patrolling. It was typical for a new border surveillance camp to engage and inflict casualties on the Viet Cong for a few weeks, then to relapse into inertia. The Viet Cong were clever enough to avoid contact on the border, as well as in the interior where they had no immediate operational interest. Their purposes were served if their main force units, groups of replacements, and resupply columns could cross such areas undetected. There was no interlocking, lateral patrol pattern between border surveillance camps. Each camp was authorized four companies. Two companies were supposed to be on the border at all times, operating from forward bases in a wide linear deployment that subdivided the company into platoons, sections, and five-man reconnaissance teams; but this arrangement did not work very well in practice. It is doubtful that the negative intelligence reports by CIDG patrols were of much value. Eighteen border sites, with a total of sixty-three strike force companies assigned, gave a density of one company to twenty-eight miles of border, or, in terms of continuous patrolling on a 24-hour basis, one platoon to twenty-eight miles. This constituted only a minor presence on the border. As a result, the border surveillance camps had no real success in controlling enemy movement across the border. The continued existence of the camps, however, was still justified, because they were able to continue valuable surveillance missions and collect a variety of intelligence data and, where located in populated areas, they contributed significantly to area development.

The U.S. Special Forces and Vietnamese Special Forces counterpart command structure had its effect on the combat operations conducted in the CIDG program. Vietnamese Special Forces counterparts consistently resisted patrolling in squads and platoons. They argued reasonably that they lacked trained, indigenous CIDG leaders. The shortage of leaders among the irregulars, however, was an inescapable consequence of the Vietnamese Special Forces refusal to allow leadership training in the camps. Even with this shortage of leaders, many border operations were planned in camp to break down into small squad-size patrols on arrival at the border, but the plan was seldom carried out. Once away from camp, the Vietnamese Special Forces patrol commanders frequently found a plausible excuse for changing the plan and keeping the force together as a company. The Vietnamese Special Forces also resisted patrolling at night. The night outpost "ambush" was a Special Forces compromise, which had the value of placing armed troops in a location outside the camp without the necessity for insisting on patrolling.

Command problems also arose. Vietnamese Special Forces men normally served as strike force commanders both in the CIDG camp and on operations. On patrol, however, the Vietnamese Special Forces often abdicated this role, with command then going to the U.S. Special Forces by default. When Vietnamese Special Forces commanders demonstrated a complete inability to exercise command in situations in which everyone's safety depended on the issuance of orders, or when delay meant defeat or allowing the enemy to escape, some U.S. Special Forces detachment commanders were virtually forced to assume command authority. Ironically, but not surprisingly, the good rapport that readily developed between Special Forces and strike force troops tended to diminish the authority of the counterpart Vietnamese Special Forces almost to the point of eliminating them from the chain of command.

In general, the operational effectiveness of the U.S. Special Forces CIDG in this period was hindered by two things. The Montagnard uprising in September 1964 caused a loss of momentum, and, combined with heavy rains in the highlands, severely restricted offensive operations for the last quarter of 1964. Until mid-1965, despite the assignment of additional missions, the pattern of CIDG operations remained essentially the same as that described above: patrols were generally of company size, and were carried out in daylight. The target was the local Viet Cong shadow government.

On 15 May 1964 the U.S. Mission in Saigon initiated a new program called Project Leaping Lena, with the mission of conducting reconnaissance operations in Vietnam. Under Leaping Lena, Special Forces A detachments trained Vietnamese Special Forces and CIDG troops in techniques of long-range reconnaissance patrolling. In June Leaping Lena began to be transferred to the Military Assistance Command and the Special Forces under Operation SWITCHBACK procedures. Under the Military Assistance Command and the Special Forces, Leaping Lena would become Project Delta, the first of the special operations which would come to be among the most powerful and effective combat operations of the Vietnam War. Project Delta had a long-range reconnaissance and intelligence gathering mission as its basic operating concept. Organized into a reconnaissance element and a reaction force, at full operating capacity Project Delta would comprise about 600 men, both U.S. and Vietnamese, plus an advisory command element organized as a modified B detachment. The typical reconnaissance element consisted of eight road patrol teams of four indigenous personnel each, and sixteen reconnaissance teams of two Special Forces and four indigenous personnel each. The reaction force was a battalion equivalent of three or more companies; for Project Delta, this force was a Vietnamese Army Ranger battalion while other projects used Montagnard troop units.

Another special operations force, the mobile strike force had its beginnings in this period. In October 1964 a mobile strike force, dubbed the "Eagle Flight," was formed in II Corps Tactical Zone. It was the forerunner of the mobile reaction forces which consisted of highly trained CIDG units organized into separate companies at Special Forces company (corps or field force) and Special Forces operational base level for use as reserve and reinforcement elements to CIDG camps threatened or under attack by superior numbers of the enemy. Mobile strike forces were also capable of conducting raids, ambushes, combat patrols, and other small-scale conventional combat operations independently, in conjunction with other CIDG units, or in support of conventional forces. When organized in 1964, the mobile strike forces were placed under unilateral U.S. Special Forces command. It would not be until December 1966 that they would come under joint U.S. Special Forces–Vietnamese Special Forces command. Throughout their existence, mobile strike forces maintained a flexibility of organization, but usually a typical company would have a headquarters element, three rifle platoons, a weapons platoon, and a reconnaissance platoon. As in the case of other special operations forces, the mobile strike forces were airborne qualified.

Viet Cong operations in the vicinity of CIDG camps continued at a high level of intensity throughout the 1963–1965 period; three major Viet Cong attacks took place and the camps at Hiep Hoa and Polei Krong were overrun. The enemy victories spotlighted two major weaknesses of the CIDG camp system at that time: each camp was on its own during the hours of darkness and if attacked the arrival of reinforcements was unlikely until daylight; and in areas where the Viet Cong effectively controlled the villagers and could rely on their silence, it was possible for a Viet Cong battalion to preserve surprise right up to the moment it launched its attack at close range. CIDG camps were also infiltrated by Viet Cong posing as loyal troopers and helping the enemy with intelligence and interior support when the attacks began.

Two nights after the Polei Krong attack, the Viet Cong attempted a similar raid on the 300-member strike force at Nam Dong in the southwest portion of Thua Thien Province in I Corps area. At 0230 on 6 July 1964 an estimated reinforced battalion of several hundred Viet Cong launched a co-ordinated attack. It opened with a shattering mortar barrage that hit most of the key installations, followed by the three ground assaults. The camp, however, was not overrun. The U.S. Special Forces troops and the Nungs put up a gallant fight and held the inner perimeter until dawn, when the Viet Cong withdrew.

Nam Dong had been scheduled to be closed out for lack of recruiting potential and its defenses were not in good shape. Tall grass had been allowed to grow on otherwise good fields of fire right up to the outer perimeter, and the defenses of the inner perimeter were reasonably sound more by chance than by a deliberately constructed defensive feature. It was an old French installation, too small to contain the strike force: it was therefore used as the core of a larger camp and housed the Special Forces and the Nungs. This inner perimeter of defense helped save Nam Dong; the "hardened" and "fighting" camps developed within the next few years would utilize the second interior perimeter defense, or mutually supporting positions, extensively and with success. Even with total penetration and breakdown of the first perimeter, a camp could be held from the inner perimeter.

There is evidence that the Viet Cong had been attempting to undermine the loyalty of the garrison. Agitators had appeared in the village and treated members of the strike force to drinks. On the day before the attack, there had been some rock-throwing; even a few weapons had been fired by the Vietnamese civilians. In the afternoon there had been a show of reconciliation, but the Nung leader feared more trouble. That night none of the Nungs slept in their quarters. All were at their posts, armed and alert, when the attack came.

A circumstance that favored the Viet Cong was a peculiar decision on the part of the district chief, whose headquarters was at Khe Tre, a village at the other end of the valley about a three-hour march away. He insisted that his Civil Guard companies do all the patrolling west of the airstrip and had forbidden the Nam Dong garrison to send any patrols into the area adjacent to the campsite in the direction from which the attack came.

The enemy approached from the northwest across the river and the airstrip. Surprise was complete. Ten mortar concentrations within fifteen minutes shattered the night and paralyzed the defenders. Only one fragmentary message was received by the B detachment at Da Nang. The radio shack was hit early, but the Special Forces radio operator who slept there rolled over as the first round exploded and tapped out at 0235 "under intense mortar attack." It is estimated that 80 percent of the casualties occurred within the first fifteen minutes of the attack when the radio shack, the Nung barracks, and the dispensary were knocked out. The camp was badly cratered—much more severely than Hiep Hoa or Polei Krong.

As the mortar barrage lifted, the first assault rolled over the outer perimeter. Strike force troops who were not actually on watch had been asleep in their barracks, but the Nungs were at their posts, manning machine guns and fighting in the communications trenches. Unlike that in Polei Krong, ammunition at Nam Dong was stockpiled close to the gun positions. The Nungs fought well; a number of them were wounded, but none were killed.

The Viet Cong had crept up to the exterior lines of barbed wire through the tall grass and breached the wires apparently before launching the attack so that they could toss hand grenades into an open 81-mm. mortar emplacement. This emplacement was the scene of heroic action when the U.S. detachment commander, Captain Roger H. C. Donlon, who was himself wounded, repeatedly attempted to rescue a team member wounded by grenades. Captain Donlon was awarded the Medal of Honor.

Two more assaults were directed at the inner perimeter. The Viet Cong charged in with great *élan*; some did not stop until disabled by the intense fire. Close examination of the dead Viet Cong after the action suggested that the attacking force had been increased by special shock troops just before the operation. Among those killed close to the inner perimeter while leading the assault elements were well set-up, well-muscled young men with close-cropped hair and clean fingernails and toenails, an indication that they had not been in the jungle very long.

The defenders held out, but dawn was welcome for they had little ammunition left. There was no relief until daylight. It was the Vietnamese Independence Day weekend and the Vietnamese pilots were off in Da Nang. Nothing was ready to go. The flareship was not even loaded, but it finally got over the target at 0430 and reported the entire camp ablaze. The Vietnamese district chief at Khe Tre had heard the firing and assembled his two Civil Guard companies for march, but, fearing ambush, he would not start until first light. As the column approached the village outside the camp, it met the last withdrawing elements of the Viet Cong.

Following the attacks on these three camps, Special Forces (Provisional), Vietnam, issued a detailed and extensive standing operating procedure on CIDG camp security for use by U.S. detachment commanders in advising their Vietnamese Special Forces counterparts. The Special Forces began to use small groups of Nungs as special security forces for certain CIDG camps, principally in I Corps and II Corps. The Nungs, many of whom had fought with the French during the Indochina War, proved to be excellent soldiers. They entered into contracts with the U.S. Special Forces and were directly responsible to them.

Logistics

In the early stages of the CIDG program, the U.S. Mission in Saigon employed a flexible and militarily unorthodox system to supply CIDG camps. The Army adopted many of these methods to provide a system of logistics unique in U.S. military history. It was not part of the U.S. Military Assistance Program for Vietnam, and it was independent of Military Assistance Advisory Group and Vietnam Army control. Funding was handled under a phase of the Operation SWITCHBACK program called PARASOL-SWITCHBACK. Under PARASOL-SWITCHBACK, funds for the CIDG program were received by Special Forces headquarters in Vietnam. The U.S. Army Counterinsurgency Support Office was established on Okinawa on 27 February 1963 and assigned the mission of supporting the Special Forces programs through triservice depots and local procurement sources and accounting for the PARASOL-SWITCHBACK funds expended outside Vietnam. Many items of clothing and equipment, for example, had to be obtained from markets in other countries because of size problems, composition of material, and equipment which had to be tailored to Montagnard measurements. Funds for procurement and salaries of civilian irregulars in Vietnam were accounted for by Special Forces headquarters.

There were ten key features of the Special Forces logistical system for the CIDG program.

1. Control of material, transportation, and funds (including CIDG troop pay) was kept in U.S. hands down to the point of issue to the ultimate users.
2. Local purchases of goods and services were authorized at all levels, with cash from current operating funds.
3. Requisitioning, justification, stock control, and other procedures were, initially at least, simple and informal.
4. Deliveries of equipment and supplies to A detachments from

higher echelons were made from the Logistical Support Center at Nha Trang directly to the A detachments.

5. Air transportation, with supplies landed or dropped, was the predominant method of delivery.

6. Maintenance by replacement took the place of repairing equipment on the site.

7. A special Counterinsurgency Support Office was established in Headquarters, U.S. Army, Ryukyu Islands, Okinawa, to control and expedite Special Forces external logistical support.

8. U.S. balance-of-payments control regulations were waived, permitting unlimited direct overseas procurement.

9. A special "quick-reacting supply and procurement" procedure was devised to provide quick procurement service in the United States for unusual needs.

10. Formal accountability was dropped on shipment of material to the Special Forces, and justification of requests was not required above the level of the Special Forces operational base.

Rations for CIDG troops, petroleum products, and other goods used on a daily basis were obtained in Vietnam through local and central procurement. Construction materials, labor, and services were also largely procured and paid for locally. Rations for American military personnel along with standard equipment and supplies were obtained from the U.S. Army Support Group and later U.S. Army, Vietnam, and the U.S. commissary in Saigon, and were supplemented by local purchases at each echelon.

The Special Forces Logistical Supply Center at Nha Trang aimed at keeping a sixty-day stock level of standard-issue items and those special items that had in effect become standard for the CIDG program, for example, uniforms and jungle boots for CIDG troops. Three of the four B detachments maintained emergency five-day stocks, particularly of ammunition and medical supplies. Because of the bad weather prevailing in I Corps during the monsoon, the forward supply point at the B detachment at Da Nang kept fifteen-day stocks. Over 9,000 line items of all classes of supplies were handled. In the last three months of 1964, a monthly average of 1,335 tons was distributed from the center at Nha Trang, of which 1,245 tons were delivered by air.

The most significant organizational change in the logistics system during this period was the establishment of forward supply points, like the one at Da Nang in I Corps and at the other B detachment headquarters in Pleiku, Saigon, and Can Tho. At each of these points a five-day emergency stock was maintained for A detachments. Four basic advantages were derived from the employment of the forward supply points in each corps tactical zone: aircraft could be used to the maximum extent, the response time to subordinate detachments' requests was reduced, the stockpile was dispersed, and a more effective use was made of sea lift.

Civic Action and Psychological Operations

The role and purpose of civic action in the context of insurgency-counterinsurgency is sometimes confused with that of psychological operations. Civic action, apart from its nation-building and humanitarian aspects, should be recognized as an activity calculated to prepare the way for psychological operations. By generating a sense of goodwill in the population through specific beneficial actions, a state of mind favorable to psychological operations is created among the people. These operations, if carefully planned and skillfully executed, promote a sense of loyalty to the government and motivate the people to co-operate with the government in order to defeat the insurgency. The two activities may proceed concurrently and may even be carried out by the same person or persons.

The border surveillance sites in isolated areas did not offer the same opportunities for civic action that existed in the area development centers. Nevertheless, Special Forces men, to the extent practicable, ran medical dispensaries, helped build schools and local markets, and initiated sanitation, agricultural, and home-improvement projects. As the CIDG program developed, it became customary for the two medical noncommissioned officers in each A detachment to hold sick call at the CIDG camp dispensary or in the adjoining village two or three times a week for the benefit of CIDG dependents, the local villagers, and any others from the surrounding countryside. After establishing this routine, the aidmen would extend their activities by conducting village sick calls in outlying villages, taking their medicines and equipment in a jeep or on their backs.

The contribution by the medical men is generally considered to be the most influential and productive of all the various civic action programs—and by far the biggest success with the people. By the spring of 1964, over 1.5 million people had been helped by the Special Forces in the medical programs under civic action. Special Forces medical men and their Vietnamese assistants treated wounds, fractures, sores, and infections; gave immunizations and pills for many diseases and illnesses; pulled teeth; and delivered babies.

TREATING A CHILD AT THE SPECIAL FORCES DISPENSARY

They also supervised and helped in the building of village dispensaries.

Co-operating with U.S. civilian agencies, the U.S. Army Special Forces conducted a wide variety of nonmedical civic actions including distribution of relief supplies to refugees (food, clothing, blankets, cooking utensils, soap, and toothbrushes—all furnished by the United States Operations Mission, CARE, religious groups, and families of Special Forces troops back home); building and repairing schools, dispensaries, playgrounds, marketplaces, pagodas, latrines, orphanages, and leprosariums; digging wells, clearing land, carrying out irrigation and drainage projects; constructing and repairing roads, bridges, and culverts; distributing tools, fertilizer, and seed received from United States Operations Mission and CARE; working for rodent and insect control; improving the grade of chickens and pigs with breeding stock provided by United States Operations Mission; building ponds and stocking them with fish provided by United States Operations Mission; distributing school books, pencils, notebooks, blackboards, and chalk; conducting classes in English for CIDG troops and in some cases local officials; and establishing co-operative stores where local

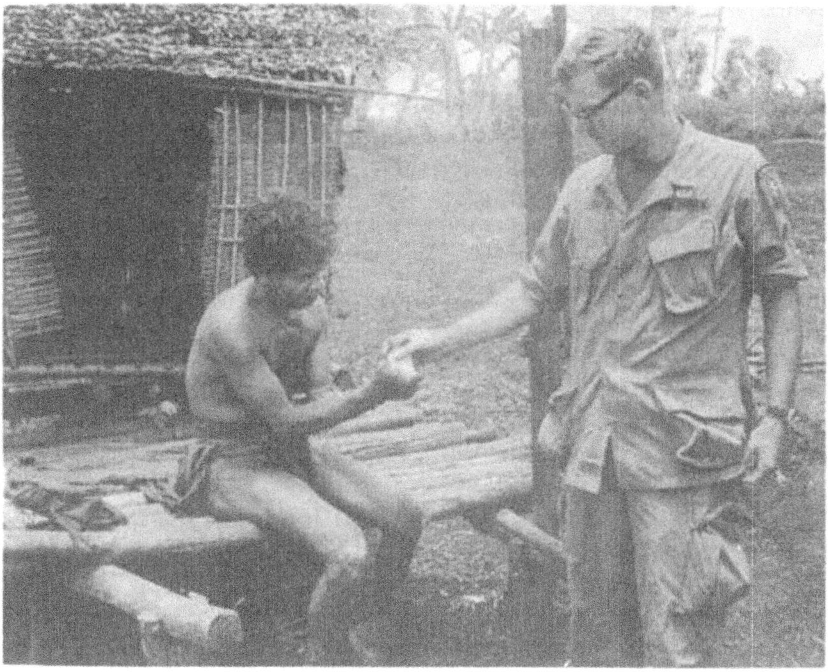

MEDICAL MAN GIVING DAPSONE PILLS TO A LEPER

produce and handicrafts could be sold and manufactured articles purchased. In remote and dangerous areas, Special Forces men sometimes acted as the forward agents or representatives for U.S. civilian agencies.

Two typical monthly reports submitted by A detachments in March 1964 further show the wide range of Special Forces activity. Civic action activities conducted by one Special Forces A detachment in one month included 925 patients treated at the camp dispensary; 1,963 patients treated in three neighboring villages; land cleared for growing dry rice and vegetables; security fence constructed for a strategic hamlet (villagers who were physically able to work furnished the labor); food provided for a leper colony nearby; two new houses constructed for the leper colony (clothing, tools, and other implements for the colony were purchased at the local market and at the district capital instead of being requested through normal supply channels, since the lepers were familiar with the local tools and worked better with them); and a co-operative store established to resell at a small profit merchandise bought at the district capital.

A recapitulation of civic action activities conducted in a single month by all Special Forces detachments deployed throughout Vietnam at that time included the distribution of 7,800 kilograms of rice, 1,000 pounds of salt, milk, fish, and other foods; 300 sets of clothing and school uniforms and quantities of yard goods and blankets; and school supplies, including textbooks, for various town schools. Twenty-four truckloads of bamboo and thatch were delivered to one village; bridges, roads, wells, and schools were repaired; cement was donated for the improvement of one pagoda and one Catholic church; 33,130 patients were treated at CIDG camp dispensaries; 23,140 patients were treated in neighboring villages by U.S. Special Forces and Vietnam Special Forces medical men or village health workers; 119 villages were visited by medical patrols; and 3,000 typhoid inoculations were administered.

Americans were the driving force behind civic action; the attitude of the Vietnamese and even the people involved, occasionally, was disappointingly indifferent. There were some encouraging instances of genuine concern on the part of district chiefs, province chiefs, Vietnamese Special Forces camp commanders, and other government officials for the welfare of the people, but many reports of Special Forces units mentioned the unsympathetic attitude of their Vietnamese Special Forces counterparts toward the civilian population. The following is an excerpt from the section on civic action in a U.S. Military Assistance Command, Vietnam, study of Tay Ninh Province dated 23 December 1963:

> Vietnamese military action in the province is almost non-existent. The scope of civic action projects being accomplished by USOM and US military personnel includes education, commerce, transportation, public welfare, health, and sanitation. But the refusal of ARVN military personnel to cooperate and assist in these projects results in failure to achieve the objectives of military civic action (goodwill leading to support for the GVN).
>
> While a few isolated instances of military civic action have occurred, the prevailing attitude among military personnel is that manual labor which assists civilians is beneath a soldier's dignity. The degree of indifference of the military toward civilian welfare was exemplified when the Vietnamese soldiers refused to help unload USOM emergency food supplies for a hamlet destroyed by the Viet Cong. The arrogant, inconsiderate treatment of civilians by soldiers has caused many civilians to support the Viet Cong.

A milder form of the military resistance to civic action was the reluctance of Vietnamese soldiers to help build houses for the civilians on the grounds that they saw no reason to help some questionable refugee to have a house when their own families' housing was so poor. Province chiefs and district chiefs were in many cases not disposed to contribute to civic action. Supplies intended for relief purposes or for civic action projects were occasionally siphoned off into commercial channels, or the intended recipients were made to pay for them instead of receiving them free.

The villagers were generally receptive to medical attention and to relief and welfare aid. The benefits to the individual were direct and tangible. Reactions to larger community-improvement projects, however, were not always what an American would expect. Villagers were willing to work on such projects if they were paid for their labor and for the materials they provided, but the concept of co-operative self-help with an unpaid contribution from each participant seemed foreign to their nature or culture patterns. Such projects had only limited success.

Psychological operations conducted in connection with the CIDG program began with emphasis on the direct day-to-day, person-to-person approach based on thorough knowledge and understanding of the ways of the local villagers and their leaders. Later, under Vietnam Army and Vietnamese Special Forces management, it became largely a mass-media program with one-time lectures to assembled groups, film showings (American westerns were very popular), loudspeaker broadcasts, and the distribution of printed matter. Until the 1964–1965 period, Vietnamese military participation in psychological operations was minimal; Special Forces provided most of the initiative, often without fixed guidelines, using their imagination in *ad hoc* programs. In the period 1964–1965 an augmentation of men qualified in civic action and psychological operations was provided to the Special Forces, with a resultant improvement in the program.

The Montagnard Uprising

The bad feeling between the Montagnards and the Vietnamese flared intoan armed uprising in September 1964. The Montagnards struck in five CIDG camps in the II Corps Tactical Zone: Buon Mi Ga, Buon Sar Pa, Bu Prang, Ban Don, and Buon Brieng. During the night of September 19–20, sixty-four CIDG troops disarmed and restricted their U.S. Special Forces advisers and rebelled against the government. At Bu Prang the mobile strike force troops killed fifteen of their Vietnamese leaders and later killed seventeen Popular Forces soldiers and two civilians at a nearby Popular Forces post. The Buon Sar Pa Mobile Strike Force with help from Bu Prang killed eleven Vietnamese Special Forces troops at their camp, seized the district headquarters at Dak Mil, and advanced on the town of Ban Me Thuot, the province capital. Two hundred Vietnamese civilians were also rounded up and held at the Buon Sar Pa camp. The CIDG force at Ban Don bound and gagged their Vietnamese Special Forces advisers and marched on Ban Me Thuot. The camp at Buon Brieng, although requested to do so, did not join in the uprising—mainly as a result of the cool-headed actions of the U.S. Special Forces detachment commander, Captain Vernon P. Gillespie.

These disorders constituted an armed challenge to the government. By late evening of 20 September, however, most of the insurgent elements had returned to their base camps. U.S. Special Forces A detachment leaders at Bu Prang, Ban Don, and Buon Brieng personally persuaded their Rhade units not to march on Ban Me Thuot, but those at Buon Sar Pa and Buon Mi Ga were less successful. The following six days were marked by an uneasy calm during which leaders of both sides negotiated. On 28 September the situation was finally brought under control by U.S. advisers acting as intermediaries.

Montagnard demands presented to Vietnamese officials at a conference in Pleiku from 15 to 17 October included: publishing of government policy in regard to landownership and relations between Montagnards and the government; representation in the National Assembly and at the district and province levels of government; positions in government bureaus and local government offices; larger Montagnard quotas in Vietnamese officer and noncommissioned officers schools; Montagnard tribunals; and certain symbols of Montagnard autonomy, for example a Montagnard flag, Montagnard languages taught in primary schools, and retention of Montagnard names for cities and villages. Government representatives agreed to some of these demands and promised to study the others.

The 5th Special Forces Group (Airborne)

In the early summer of 1964, two decisions were made concerning the future employment of the U.S. Special Forces in Vietnam: to expand the Special Forces organization to a full group, and to change the tour from a six-month temporary duty assignment to a one-year permanent change of station.

The emphasis on the border surveillance mission after the takeover from the U.S. Mission, the further development of that mission as stated in the Vietnamese Pacification Plan for 1964, and other new missions under consideration created a requirement for

TABLE 3—ORGANIZATION OF U.S. SPECIAL FORCES DETACHMENTS

Operational Detachment C	
Commanding officer, lieutenant colonel	1
Executive officer, major	1
Adjutant, captain	1
Intelligence officer (U.S. Army), captain	1
Operation and training officer (U.S. Army), captain	1
Supply officer (U.S. Army, captain	1
Sergeant major E9	1
Intelligence sergeant E8	1
Operations sergeant E8	1
Supply sergeant E8	1
Assistant supply sergeant E7	1
Radio operator supervisor E7	1
Administration supervisor E6	1
Chief of research and development operators E5	4
Senior field radio repairman E5	1
Operational clerk E5	1
Operational Detachment B	
Commanding officer, major	1
Executive officer, captain	1
Adjutant, captain	1
Intelligence officer (U.S. Army), captain	1
Operation and training officer (U.S. Army), captain	1
Supply officer (U.S. Army), captain	1
Sergeant major E9	1
Intelligence sergeant E8	1
Operations sergeant E8	1
Chief medical noncommissioned officer E8	1
Heavy weapons leader E7	1
Light weapons leader E7	1
Engineer sergeant E7	1
Radio operator supervisor E7	1
Supply sergeant E7	1
Administration supervisor E6	1
Assistant supply sergeant E6	1
Preventive medicine specialist E6	1
Chief of research and development operators E5	4
Engineer E5	1
Operational clerk E5	1
Operational Detachment A	
Commanding officer, captain	1
Executive officer, lieutenant	1
Operations sergeant E8	1
Heavy weapons leader E7	1
Intelligence sergeant E7	1
Light weapons leader E7	1
Medical specialist E7	1
Radio operator supervisor E7	1

TABLE 3—ORGANIZATION OF U.S. SPECIAL FORCES DETACHMENTS—Continued

Operational Detachment A—continued	
Engineer sergeant E7	1
Assistant medical specialist E6	1
Chief of research and development operator E5	1
Engineer E5	1
Administrative Detachment	
Commanding officer, captain	1
Executive officer, lieutenant	1
Detachment sergeant E7	1
Mess steward E7	1
Supply sergeant E6	1
First cook E6	1
First cooks E5	2
Armorer E4	1
Detachment clerk E4	1
Cooks E4	3
Equipment reports clerk E4	1
Cooks apprentice E3	1

TABLE 4—STATUS REPORT, U.S. SPECIAL FORCES CIDG CAMPS, OCTOBER 1964

Location	Vietnamese Special Forces		U.S. Special Forces		Strike Force		Nung Force
	Officers	Enlisted Men	Officers	Enlisted Men	Companies	Men	Men
I Corps Tactical Zone							
Da Nang	6	14	7	16	0	53	158
Khe Sanh	2	11	2	10	5	677	48
Kham Duc*	1	12	2	10	2	183	118
	1	12	2	10	5	507	408
A Ro	1	13	2	10	4	338	53
A Shau	4	10	2	10	3	332	45
Ta Ko	1	6	1	10	2	231	0
II Corps Tactical Zone							
Pleiku	14	54	12	29	0	0	60
Kannack	5	17	2	10	3	418	36
Gia Vuc	2	8	2	10	3	541	74
Dak Pek	1	16	2	10	4	479	33
Dak To	1	10	2	10	3	371	—
Buon Beng	2	10	2	10	5	865	—
Polei Krong	1	10	2	9	4	463	25
Plei Mrong	3	13	2	10	3	464	37
Chu Dron	1	8	2	5	4	444	35

TABLE 4—STATUS REPORT, U.S. SPECIAL FORCES CIDG CAMPS, OCTOBER 1964—Continued

Location	Vietnamese Special Forces		U.S. Special Forces		Strike Force		Nung Force
	Officers	Enlisted Men	Officers	Enlisted Men	Companies	Men	Men
II Corps Tactical Zone—continued							
Plei Do Lim	2	9	1	5	2	271	41
Plei Me	1	10	1	5	3	241	35
Phey Srunh	1	3	2	10	4	652	0
Buon Mi Ga	1	3	1	5	4	619	0
Nha Trang	5	11	1	5	3	505	0
Ban Don	2	12	2	10	5	799	0
Plei Ta Nangle	1	11	2	10	4	606	0
Dong Tre	1	9	2	10	4	619	51
Dong Ba Thin	—	14	1	5	2	308	9
Buon Brieng	1	10	2	10	3	775	22
Plei Djereng	1	1	1	5	2	231	30
Bu Prang	0	0	2	10	[b]	[b]	[b]
III Corps Tactical Zone							
Saigon	8	13	6	17	0	0	0
Bu Chia Map	4	10	2	10	4	449	9
Suoi Da	5	8	2	10	5	570	0
Bu Dop	3	8	2	10	3	275	0
Minh Thanh	4	8	2	10	3	382	0
Trang Sup	4	15	1	5	7	856	1
Nui Ba Den	[b]	[b]	1	5	2	175	0
Loc Ninh	3	7	2	9	4	596	0
Long Thanh	[b]	[b]	3	12	[b]	[b]	[b]
Thu Duc	3	15	2	9	[b]	[b]	521
Ben Cat	[b]	[b]	1	5	4	450	[b]
IV Corps Tactical Zone							
Can Tho	[c]70		10	17	83	—	0
Moc Hoa	[c]23		2	10	4	608	33
Don Phuoc	[c]10		2	10	2	462	0
An Long	[c]11		2	10	3	434	0
An Phu	[c]13		2	10	5	754	0
Tinh Bien	[c]23		2	10	4	632	0
Vinh Gia	[c]18		2	10	3	436	1
To Chau	[c]13		2	10	3	392	0

[a] Two detachments.
[b] Not available.
[c] Combined figures for officers and enlisted men.

TABLE 5—U.S. SPECIAL FORCES CIDG CAMPS ESTABLISHED IN VIETNAM, JULY 1961–OCTOBER 1964

Campsite	Province	Date of Opening	Mission[a]	Date of Closeout or Turnover	Remarks
I CTZ (as of Oct 64)					
An Diem	Quang Nam	Mar 63	CIDG	Jul 64	
A Ro	Quang Nam	Jun 64	BS/CIDG		
A Shau	Thua Thien	Mar 64	BS/CIDG		Lowland Vietnamese in area
Ba To	Quang Ngai	Nov 62	CIDG	Feb 63	
Da Nang	Quang Nam	Sep 62	Control		B detachment
Gia Vuc	Quang Ngai	Feb 63	CIDG		Rhe tribal group in area
Hiep Duc	Quang Tin	Apr 63	CIDG	Dec 63	
Hoa Cam	Quang Nam	Nov 61	CSD Training Center	May 63	Reopened temporarily in mid-1964 as preliminary recruit training center prior to deployment of recruits to Khe Sanh and Kham Duc (6-wk training without weapons)
Kham Duc (2)	Quang Tin	Sep 63, Feb 64	CIDG/BS		Lowland Vietnamese in area; first camp apparently closed out and reopened
Khe Sanh	Quang Tri	Aug 62	CIDG, TW; later BS		Bru, lowland Vietnamese in area
Phu Hoa	Quang Nam	31 Dec 62[b]	CIDG	Jun 63[b]	
Son Ha (Ha Thanh)	Quang Ngai	Jan 63	CIDG	Feb 64	
Ta Bat	Thua Thien	Mar 63	CIDG	Mar 64	
Ta Ko	Quang Nam	Sep 64	BS/CIDG		Lowland Vietnamese in area
Ta Rau (Nam Dong)	Thua Thien	Mar 63	CIDG	Jul 64	Resettled Vietnamese in area
Tra Bong	Quang Ngai	By Dec 62	CIDG, MC	Aug 63	
Tra My	Quang Tin	Oct 62	CIDG	Feb 64	

II CTZ (as of Oct 64)					
Ban Don	Darlac	Dec 62	CIDG/BS		Rhade and Jarai tribal groups in area; part of Buon Enao complex
Bon Sar Pa	Quang Duc	Jun 63	CIDG	Sep 64	Mnong and Rhade tribal groups in area
Buon Brieng	Darlac	Mar 64	CIDG (GWOA)		Rhade tribal group in area
Buon Dan Bak	Darlac	Dec 62[b]	CIDG	Jan 63	Part of Buon Enao complex
Buon Enao complex	Darlac	Nov 61	CIDG	Jun 63	Rhade tribal group in area
Buon Bi Ga	Darlac	Dec 62	CIDG Training Center, later CIDG (GWOA)		Mnong and Rhade tribal groups in area
Buon Uing	Darlac	Apr 63	CIDG/BS	Jul 64	Rhade, Jarai, and Mnong tribal groups in area
Buon Yun	Darlac	May 63	CIDG/BS	May 64	Rhade, Jarai, and Mnong tribal groups in area
Bu Prang	Quang Duc	May 63	CIDG/BS		Mnong tribal group in area
Cheo Reo (Buon Beng)	Phu Bon	May 63	CIDG	Aug 63	Bahnar and Jarai tribal groups in area
Chu Dron (Duc Co)	Pleiku	May 63	CIDG/BS		Bahnar and Jarai tribal groups in area
Cung Son	Phu Yen	Jun 62	CIDG		
Dak Pek	Kontum	Nov 62	CIDG/BS		Sedang tribal group in area
Dak To	Kontum	Feb 63	CIDG/BS		Sedang tribal group in area
Dam Pau (Dalat Serignac, Camp Brotherhood)	Tuyen Duc	Aug 62	CIDG	Jan 64	Koho tribal group in area
Djirai	Lam Dong	Mar 63	CIDG	Mar 64	Koho, Bulach tribal groups in area
Dong Ba Thin	Khanh Hoa	Mar 64	CIDG/VNSF Training Center		1 B and 1 A detachment
Dong Tre	Phu Yen	Jul 63	CIDG		
Kannack	Binh Dinh	Feb 64	CIDG (GWOA)		Rhe tribal group in area
Krong Kno	Darlac	Nov 62	CIDG	Jun 63	Moang, Koho, and Rhade tribal groups in area

TABLE 5—U.S. SPECIAL FORCES CIDG CAMPS ESTABLISHED IN VIETNAM, JULY 1961–OCTOBER 1964—Continued

Campsite	Province	Date of Opening	Mission[a]	Date of Closeout or Turnover	Remarks
Mang Buk	Kontum	May 62	CIDG	Aug 63	Katang tribal group in area
Nha Trang	Khanh Hoa	Oct 62[b]	Control [S1, S4 Hq, USASF(P)V], later A detachment to advise VNSF		Cham, Rhade tribal groups in area
Phan Rang (Phuoc Thien)	Ninh Thuan	May 62	CIDG	Jul 63	Cham tribal group in area
Phey Srunh	Tuyen Duc	Jan 64	CIDG (GWOA)		Koho tribal group in area
Plei Djereng	Pleiku	Jun 64	CIDG/BS		Jarai and Rhade tribal groups in area
Plei Do Lim	Pleiku	Apr 62	CIDG		Jarai tribal group in area
Pleiku	Pleiku	Oct 62	CIDG Montagnard Training Center, later Control		Jarai tribal group in area; B detachment arrived by Dec 62
Plei Me	Pleiku	Oct 63	CIDG/BS		Jarai tribal group in area
Plei Mrong	Pleiku	Nov 62	CIDG/BS		Jarai tribal group in area
Plei Ta Nangle	Binh Dinh	Dec 63	CIDG		Bahnar and Rhade in area
Plei Yt	Pleiku	31 Dec 62[b]	MS	Jun 63	
Polei Krong	Kontum	Feb 63	CIDG/BS		Bahnar, Jarai, Sedang, and Mnong tribal groups in area
Song Mao	Binh Thuan	May 62	Training center for 77th SFG, CIDG	Mar 64	
Tan Canh	Kontum	Aug 62[b]	CIDG and TW	Feb 63	
Van Canh	Binh Dinh	Nov 62	CIDG	Aug 63	Jarai tribal group in area
III CTZ (as of Oct 64)					
Ben Cat	Binh Duong	Sep 64	Road security, inter-		

			diction and counter-guerrilla operations		
Ben Phuoc	Long An	Mar 64	CA PSYWAR	Moved to Ben Cat Nov 64	Under USMACV control
Bien Hoa	Binh Hoa	Oct 64	Control		Part of B detachment shared with Saigon
Bu Dop	Phuoc Long	Dec 63	BS/CIDG		Stieng tribal group in area
Bu Ghia Map	Phuoc Long	Jun 63[b]	CIDG/BS	May 64[b]	Stieng, Vietnamese, and Cambodians in area
Go Dau Ha	Tay Ninh	Apr 64	BS-FOB for Trang Sup	Aug 64	
Hiep Hoa	Long An	Feb 63	CIDG/BS	Jan 64	Anticipated turnover to province 31 Dec 63 as SDC Training Center
Los Ninh	Binh Long	Mar 63	CIDG/BS		Cambodians in area
Long Thanh	Bien Hoa	Jul 61	Airborne Ranger Training Center		
Minh Thanh	Binh Long	Nov 63	CIDG/BS		Vietnamese and Cambodians in area
Nuoc Vang	Phuoc Thanh	Feb 63	MS and CIDG	Jan 64	Cambodians, Montagnards, and resettled Vietnamese in area
Nui Ba Den	Tay Ninh	Aug 64	Radio Relay Station		Vietnamese and Cambodians in area
Saigon	Gia Dinh	Sep 62	Control		First C detachment arrived Sep 62 to function as temporary USASF Hq
Suoi Da	Tay Ninh	Jun 64	CIDG-BS		Vietnamese and Cambodians in area
Thu Duc	Gia Dinh	Aug 62[b]	CSD Training Center similar to Hoa Cam; Later, classified mission		Vietnamese and Nungs in area; sources indicate camp may have been closed at one time and reopened in Jul 64
Trang Sup	Tay Ninh	Feb 63	CIDG and MS, Later BS		Cao Dai and Cambodians in area

TABLE 5—U.S. SPECIAL FORCES CIDG CAMPS ESTABLISHED IN VIETNAM, JULY 1961–OCTOBER 1964—Continued

Campsite	Province	Date of Opening	Mission[a]	Date of Closeout or Turnover	Remarks
Tuc Trung	Long Khanh	Feb 63	CIDG and MS	Dec 63	Bulach and Chrao tribal groups in area
IV CTZ (as of Oct 64)					
An Long	Kien Phong	Jan 63	CIDG/BS		Vietnamese, Cambodians, and Hoa Hao in area; initial work with Catholic Youth
An Phu	An Giang	Apr 64	BS/CIDG		Hoa Hao in area
Binh Thanh	Kien Tuong	Oct 64	CIDG/BS		Vietnamese, Cambodians, and Hoa Hao in area
Can Tho	Phong Dinh	Jan 63	Control		B detachment
Chau Long	An Giang	Oct 62	CIDG	Apr 64	Work primarily with Catholic Youth
Don Phuoc	Kien Tuong	Aug 64	BS/CIDG		Vietnamese and Hoa Hao in area; opened initially in Jun 64 as FOB
Ha Tien (To Chau) ..	Kien Giang	Feb 63	BS/CIDG		Vietnamese and Cambodians in area
Long Khanh	Vinh Binh	Feb 63	CIDG	Jul 64	Vietnamese in area; work primarily with Catholic Youth
Long Phu	Chuong Thien	Feb 63	CIDG/BS	Feb 64	Work primarily with Catholic Youth
Luong Tam	Chuong Thien	Dec 63	CIDG/BS	Jun 64	Vietnamese in area; initially FOB for Long Phu
Moc Hoa	Kieng Tuong	Jan 64	CIDG/BS		Vietnamese and Cambodians in area
Soc Trang (Du Tho) ..	Ba Xuyen	Aug 62	CIDG	Jul 63	CIDG troops not released until Apr 64; work primarily with Catholic Youth

Tan Hiep	Kinh Tuong	Mar 64	Special MAAG mission	Moved to Ben Phuoc May–Jun 64	
Tan Phu	An Xuyen	Apr 63	CIDG	Jun 64	
Tinh Bien	An Giang	Apr 64	BS/CIDG		Vietnamese and Cambodians in area
Vinh Gia	An Giang	Aug 64	BS/CIDG		Vietnamese and Cambodians in area
Vinh Loi	Go Cong	Mar 64	CAFF/PSYOP under USMACV control		

[a] CIDG—Civilian Irregular Defense Group (area development); BS—border surveillance; TW—Trailwatcher; MC—mountain commando; GWOA—guerrilla warfare operational area (enemy-controlled area to be penetrated and recovered by guerrilla forces); MS—mountain scout; CAFF—civil affairs; FOB—forward operational base.

[b] Deadline date.

more Special Forces detachments. In March 1964, the commanding officer of Special Forces (Provisional), Vietnam, had requested eleven additional A detachments, three B detachments, and four C detachments—in effect an increase to a full group. In the spring of 1964, Special Forces (Provisional), Vietnam, became an established command and was officially designated U.S. Army Special Forces, Vietnam. Finally, on 1 October 1964, the 5th Special Forces Group (Airborne), 1st Special Forces, with an authorized strength of 1,297, was assigned to Vietnam. The introduction of the 5th Special Forces Group and the transition to permanent change of station was to be phased, with the first nine permanent change of station A detachments arriving on 28 September 1964. All temporary duty detachments were to depart by 1 May 1965. By the end of September there were five B and forty-four A detachments in Vietnam, of which one B and three A detachments were assigned to special projects, plus a signal element and twenty-man psychological operations and civic action augmentation unit. By February 1965 Special Forces strength would rise to four C, twelve B, and forty-eight A detachments. (*Tables 3 and 4*)

With the expansion to group level, C detachments were to replace B detachments at the corps level and B detachments were to continue to control A detachments and possibly were to be collated with Vietnamese Army division headquarters. Headquarters, 5th Special Forces Group (Airborne), was to exercise command less operational control of deployed detachments, which would remain for the time being with the U.S. senior advisers in each corps; to advise on opening and closing of camps; to establish new camps; and to advise the Vietnamese Special Forces High Command. (*Table 5*) The 5th Special Forces Group would also provide formal training when required for Vietnamese Special Forces and CIDG units.

PART TWO

THE MIDDLE YEARS: 1965–1968

CHAPTER IV

The CIDG Program Begins To Mature

In the early years the Civilian Irregular Defense Group program was essentially a defensive effort characterized by the overriding goal of securing control over the indigenous minorities and winning their allegiance so that they would not fall to the Communists. The missions were to control the Viet Cong, either through area development or border surveillance or combinations of the two. The civilian irregulars and the U.S. Special Forces were not hunting the Viet Cong in the beginning. The buildup of conventional U.S. forces in Vietnam changed all that and opened the door to the next stage in the evolution of the Special Forces CIDG program—a stage in which the Special Forces and the irregulars would find themselves cast in a distinctly offensive role. They were to become hunters with the mission of finding and destroying the enemy.

In January 1965, just before the beginning of the massive U.S. commitment of conventional forces to South Vietnam, the U.S. Special Forces counterinsurgency program was defined in a letter from Headquarters, 5th Special Forces Group (Airborne), to the commanders of all operational A, B, and C detachments:

> *Definition*: The SF Counterinsurgency Program is a phased and combined military-civil counterinsurgency effort designed to accomplish the following objectives: (a) destroy the Viet Cong and create a secure environment; (b) establish firm governmental control over the population; and (c) enlist the population's active and willing support of, and participation in the government's programs.
>
> These objectives are accomplished while executing any one of three possible assigned missions: (1) border surveillance and control, (2) operations against infiltration routes, or (3) operations against VC war zones and bases.
>
> *Concept of the Operation*: This is essentially a clear, secure, and develop operation. A fundamental point in the counterinsurgency program is that, where possible, the Strike Force personnel should be locally recruited in order to provide an exploitable entry to the populace which, in turn, facilitates military-civil relations.

The letter goes on to state that no population area which is "uncommitted" or which has been dominated by the Viet Cong can be

AT EASE IN CAMP, MONTAGNARDS ARE AMUSED *at American's attempts to play the native flute (left) and the ritual of shaving (right).*

considered won to the government until the Viet Cong have been cleared from the area, the local Viet Cong underground organization has been eliminated, and the government of the Republic of Vietnam has firmly replaced that of the Viet Cong. The letter also points out that in remote areas the task is often to introduce the government representatives for the first time.

While the new offensive role of the CIDG under the U.S. Special Forces is reflected in the letter, its operations statement reveals that the old area development concepts were still operative to a large extent. As the number of conventional U.S. forces began to grow, however, the use of the Special Forces and CIDG troops in a straightforward, offensive combat role became the norm—both in theory and in practice. In these middle years, the civilian irregulars under the Special Forces assumed a fully offensive, though not always fully conventional, role. They became hunters of the Viet Cong and the North Vietnamese Army.

U.S. forces were there to defeat the enemy. Their presence and the presence of conventional North Vietnamese units changed the nature of the entire war. Before 1965 there was principally a guerrilla insurgency. After 1965 the conflict became more conventional, with the major qualification that guerrilla tactics were used heavily by the enemy. At any rate, the "conventionalization" of

STRIKE FORCE ASSAULT TROOPS LEAVING HELICOPTER

the war led to the "conventionalization" of the civilian irregulars, who were no longer fighting for their own protection but instead were fighting to defeat the enemy.

The emergence of the Special Forces Civilian Irregular Defense Group project as an offensive effort is not surprising. For the most part, the American soldier arriving in Vietnam found himself in an environment totally different from anything he had ever experienced. He was not used to the heat, the rain, the jungle; he did not know the Vietnamese people and their culture; he did not speak the language; and, most significantly, he did not know who or where the enemy was or how to find out. On the other hand, the U.S. Special Forces and their civilian irregular troops were accustomed to the heat, rain, and jungle, and they could communicate. Special Forces men had come to know the people and their culture. Participation by the Special Forces in tribal ceremonial functions was not uncommon, and the Montagnard bracelet worn by many Green Berets was a token of Montagnard respect and involved a ceremony for its presentation. And, finally, if they did not know who or where the Viet Cong were, they could find out.

Besides causing the shift to the offensive in the Special Forces

CIDG role, the large number of American troops had other effects on the CIDG program, both good and bad. For example, new American commanders often misjudged or misunderstood the capabilities of the Civilian Irregular Defense Group units present in their tactical areas of responsibility. As a result a company of irregulars would occasionally be requested for a job which a U.S. infantry company could handle but for which the irregulars were neither trained nor prepared. Another mistake U.S. commanders made was to propose splitting a CIDG company, with the idea ol sending one platoon here and another there. This practice, standard procedure for an American unit, was hard on the CIDG troops for whom unit integrity was extremely important. The civilian irregulars did not think of themselves as battalions, brigades, or divisions. They were companies, strike forces, tied together not only by their Vietnamese Special Forces commanders and U.S. Special Forces advisers, but also by their common homelands and tribal bonds. It is not surprising then that a CIDG platoon would be in over its head trying to work with a U.S. infantry platoon on its flanks. On the whole, U.S. commanders never really became familiar with the civilian irregulars and their capabilities.

The U.S. buildup also had good effects. Most important of these was in the area of camp defense and security—two major concerns in the early years. The chief difficulty had been to reinforce a camp rapidly and effectively when it came under imminent or actual enemy attack. The buildup of U.S. forces not only provided powerful U.S. reaction forces, but it also promoted further development, particularly from 1966 to 1967, of indigenous reaction forces, known as mobile strike units, in numbers, dispersion, and strength. The advent of U.S. Air Force support along with the helicopter and its availability to the CIDG reaction forces made rapid and effective response a reality. Command and operational control structures were reorganized and streamlined in order to provide rapid reaction forces with reserves. Eventually, these developments would make it possible for the Special Forces and civilian irregulars to reinforce camps under attack that would otherwise have been lost. Similar positive effects were achieved in other combat operations. For example, in the early 1960s the irregulars had run into enemy units which were too big for them to handle. In the latter years, such contacts could be and were exploited. In fact, one of the major functions of the CIDG came to be precisely that of finding the enemy in force so that he could be engaged.

Along these lines, it is not surprising that the ability to gather intelligence that was inherent in the Special Forces CIDG effort came to be heavily used. The CIDG troops and their U.S. and Vietnamese Special Forces leaders were ideally suited for the task of finding and fixing enemy forces. Their camps were dispersed from one end of the country to the other, usually in Viet Cong territory. Further, the civilian irregulars and the Special Forces were themselves trained guerrillas, capable of meeting the enemy on his own terms. Military Assistance Command, Vietnam, estimates indicate that in the course of this period almost 50 percent of the command's ground combat intelligence came from Special Forces and civilian irregulars.

Again, this new role as intelligence gatherers had both good and bad effects on the Special Forces CIDG program. On the positive side, the CIDG program was revitalized and strengthened when its value as a source of intelligence was realized and exploited. In the early years plans had been formulated for the discontinuance of the program as such, and its integration into the conventional Vietnamese military structure. In fact, the opposite happened. The program continued to expand vigorously as new missions like the production of intelligence were devised. (*Table 6*)

On the other hand, the emphasis on producing reliable intelligence for use by conventional forces necessarily led to the decline of what may be termed "local" intelligence. Under the area development concept, emphasis had been placed on intelligence covering the local Viet Cong underground organization in the area of operation of each camp. The aim was to destroy the Viet Cong organization in each tactical area of responsibility. Later, when priority was given to intelligence for conventional forces, local intelligence efforts deteriorated, with a corresponding decrease in the effectiveness of local area development. Finding and destroying the Viet Cong countrywide by conventional military methods took precedence over the more subtle tactic of systematically rooting out the Viet Cong structure in each of the areas surrounding the CIDG camps. This is not to say that the expanded and unified intelligence mission given the Special Forces and civilian irregulars was not effective. On the contrary, it was effective. But the talents of the Special Forces in area development came to be exploited only to a minimal degree.

Finally, the demand for intelligence was the primary factor influencing the development of the so-called unconventional operations carried out by the Special Forces during this period. The mission of finding the enemy led to the establishment of such special operations as Projects Delta, Omega, and Sigma, which

TABLE 6—EXPANSION OF 5TH SPECIAL FORCES GROUP AND SPAN OF ADVISORY OPERATIONS, OCTOBER 1964–JULY 1967

Date (End of Month)	Assigned Group Strength[a]	Number of Detachments			Number of Detachments With Advisory Mission		CIDG Strength		Regional and Popular Forces Strength	Total Indigenous Paramilitary Advised
		C	B	A	Sector	Subsector[b]	Camp Strike Force	Mobile Strike Force[c]		
Oct 64	951	0	5	44	0	0	19,000	[d]	0	19,000
Dec 64	1,227	4	6	48	0	1	20,400	1,600	600	22,600
Mar 65	1,338	4	11	48	0	7	19,000	1,150	4,500	24,650
Jun 65	[d]	4	11	54	5	31	22,000	1,450	18,000	41,450
Sep 65	[d]	4	11	61	5	38	24,400	1,900	25,000	51,300
Dec 65	1,828	4	11	62	5	36	30,400	1,800	28,800	61,000
Apr 66	2,318	4	14	74	7	37	34,300	2,100	36,000	72,400
Jul 66	2,627	5	16	76	7	31	33,400	2,400	34,000	69,800
Oct 66	2,589	5	16	76	6	32	34,800	3,200	28,000	66,000
Jan 67	2,745	5	16	80	6	29	[d]	[d]	[d]	[d]
Apr 67	2,726	5	16	80	6	24	32,350	5,500	24,750	62,600
Jul 67	2,657	5	16	80	4	23	34,350	5,700	18,250	58,300

[a] Strength figures are only approximate. Official 5th Special Forces Group reports contain numerous irreconcilable discrepancies, for instance, between the number of detachments with sector and subsector mission and total of Regional Forces and Popular Forces advised for September and December 1965.

[b] Included in totals of previous column. B detachments had sector mission and A detachments had subsector, but the subsector column includes some B detachments with both sector and subsector missions.

[c] Or equivalents.

[d] Not available.

proved to be significant contributions of the U.S. Special Forces to the war in Vietnam. The intelligence these projects produced was invaluable, and their deadly effectiveness against the enemy proved their worth as methods of offensive counterinsurgency.

In the course of these middle years, the Special Forces troops also were given new missions, which, while related to the CIDG program, were in addition to their CIDG mission. Included among these were the subsector advisory mission and the Recondo (reconnaissance-commando) School mission.

Combat actions during this period fell into three categories. First, there were actions connected with the opening, closing, or defense of CIDG camps, especially along the border, including the battles at A Shau, Lang Vei, Con Thien, Loc Ninh, Thuong Thoi, and Bu Dop. The pattern seems to have been that when a camp became a real nuisance to the enemy he was very likely to attack in great strength in an effort to overrun and destroy it. Otherwise the camps were left alone, and any contacts made were, for the most part, the result of sending out patrols from the camp.

Second, there were combat actions that grew out of the special operations, including Project Delta operations in I Corps; BLACKJACK 33 operations in III Corps under Project Sigma—the first operation in which mobile guerrilla forces were employed in conjunction with a project force; BLACKJACK 41, in which two mobile strike force companies conducted a parachute assault in the Seven Mountains region of IV Corps, and a mobile strike force carried out an operation in III Corps around Soui Da, in which the force was credited with rendering a Viet Cong battalion ineffective and which eventually developed into Operation ATTLEBORO.

Finally, the third type of combat action took place when CIDG troops were employed in conjunction with conventional forces in conventional combat operations. Among these were Operation NATHAN HALE, jointly conducted by CIDG forces, 1st Cavalry Division (Airmobile), and 101st Airborne Division; Operations HENRY CLAY and THAYER; Operation RIO BLANCO in I Corps, involving CIDG troops, Regional Forces, Vietnam Army troops, Vietnam Rangers, Korean marines, and U.S. marines; and Operation SAM HOUSTON in II Corps, conducted by the 4th Infantry Division and the CIDG troops. The CIDG troops also fought in the cities during the enemy *Tet* offensive of 1968. Contacts made by mobile strike forces and mobile guerrilla forces often developed into significant combat actions.

The introduction of large numbers of U.S. forces to the conflict in Vietnam brought drastic changes in the role of the U.S. Special

Forces, in the Civilian Irregular Defense Group program, and in the war itself. The effects were felt almost immediately in the assignment of new subordinate missions, including that of assisting in the introduction of U.S. forces into remote areas. Initially the U.S. troop buildup had its greatest impact on the CIDG program in the II and III Corps Tactical Zones because conventional operations were, in the beginning, conducted there on a much larger scale geographically than they were in IV Corps or I Corps. In II and III Corps, tactical areas of responsibility of the CIDG camps repeatedly intersected or were included in the operational areas of conventional U.S. forces, with the result that camp strike forces came more and more under indirect U.S. operational control.

U.S. unit commanders soon realized that a Civilian Irregular Defense camp was an excellent source of local information. They learned quickly that guides, interpreters, scouts, or trackers, and fairly proficient prisoner of war interrogators were to be found in the camps, and that the strike force companies—provided that their special aptitudes were exploited and they were not expected to perform in all respects like U.S. infantry companies—could be useful adjuncts to U.S. search and destroy operations. In I Corps conventional operations by the III Marine Amphibious Force were concentrated near the coast and therefore did not become operationally involved with the Civilian Irregular Defense Group. It was not until the spring of 1967, with the introduction of Task Force OREGON, that this situation changed. In IV Corps the U.S. buildup had no impact on operations in the beginning, although good effects were felt, for example, in more helicopter and tactical air support.

The buildup benefited the Civilian Irregular Defense Group program itself in many ways. U.S. Army engineers were brought in to assist in the construction of camps. U.S. ground forces could now be used as security forces when areas were being explored to select new campsites. A combination of U.S. combat forces and civilian irregulars would make it possible to establish CIDG camps in areas where enemy strength had previously made it unfeasible. There was now increased helicopter support for CIDG airmobile operations. U.S. combat forces could now be employed as reaction forces to exploit opportunities developed by CIDG operations and to relieve CIDG troops or camps under attack. The camps of the civilian irregulars were small, isolated strongpoints, without any inherent capability for mutual support. The organic mobile strike forces did not attain significant strength until late in 1966; Vietnamese Army reaction forces were usually available but could seldom be committed quickly. Actually, until the U.S. infantry arrived in strength there was no force available to exploit such a target of opportunity as a multibattalion concentration of Viet Cong preparing to attack a Special Forces camp, or to justify penetration of the war zones by CIDG reconnaissance patrols to locate enemy units. Especially during 1965 and 1966 many of the more productive U.S. operations in II and III Corps began as reactions to CIDG contacts with the enemy or as attacks against enemy concentrations discovered in their preparations by Special Forces agencies.

The troop buildup also brought difficulties for the Special Forces. These invariably stemmed from a lack of understanding on the part of U.S. commanders at all levels of the nature of the CIDG program and its command structure, of the role of Special Forces operational detachments, and of the capabilities and limitations of irregulars. The most common mistakes of U.S. commanders and staff officers were to equate a strike force company with a regular infantry company and to assume that CIDG camps located in an American unit's assigned area automatically came under its operational command.

In the period from July 1965 to June 1966 the Special Forces continued to grow. No attempts at turnover were made during this year; the emphasis was on expansion and development of CIDG resources in support of the war effort in general. This growth was another effect of the U.S. buildup, one which had not been expected and which came to light when the 5th Special Forces Group submitted its analysis of the CIDG.

> With the deployment of conventional U.S. combat units to RVN in May 1965, it was felt by many that the CIDG effort was no longer required or valid in the face of the increased enemy threat. A study was conducted by the MACV staff to determine the desirability of completely phasing out all USASF and converting the CIDG to Regional Force status by 1 January 1967. The proposed conversion schedule required a specified number of CIDG to convert to RF during a specified time frame.
>
> While being considered for complete phase-out by January 1967, it was found that the CIDG camps provided valuable staging bases from which ARVN and the Free World Military Assistance Forces (FWMAF) could launch offensive operations against the enemy, a role which the founders of the program had not foreseen.

Some of the points made by the departing commander of the 5th Special Forces (Airborne) in June 1966 concerning his year of command will bear repeating here.

SPECIAL FORCES MEN AND CIVILIAN IRREGULARS *on patrol pause to eat.*

The "special" about Special Forces is simply that the non-commissioned officers are the finest to be found anywhere in the world. Their multiple skills and individual motivation are exploited to the fullest in the combat environment of the A detachment in VC-dominated areas. If today's Special Forces NCO has ever had any peer, it was probably the tough, self-reliant, combat-tested soldier who fought on the Indian frontier of our own country during the 1870's. . . .

Also, to my surprise, I discovered that the CIDG troops are not the band of unskilled, disorganized, and disgruntled peasants I had envisioned. They are, in fact, closely knit religious or ethnic minority groups with a fierce loyalty to each other and to those who will treat them with respect and consideration. . . .

The addition of an airmobile company (light) has proved to be of particular value. Command and control are greatly enhanced, and the organic capability to stage my own airmobile operations has drastically reduced reaction time, which is so important in this counterinsurgency environment. Similarly, the engineer augmentation has greatly assisted in improving airfields and expediting camp construction. . . .

A few observations should be made in the area of combat operations. First, much of the initiative still rests in the hands of the VC and the NVA. This remains so primarily as a result of the first-rate intelligence system of the VC. In general, most engagements occur when the VC determine that it is to their advantage to fight. Otherwise, they fade into the bush. Operations of larger-than-company size seldom make contact. Small unit patrolling and ambushes (squad and platoon), backed up by responsive air support and reaction forces, have proven to be a most effective approach. . . . In this manner we have been able to average 9.7 contacts per day with the enemy during the past year. Counting operations outside the camp perimeters each day during the past twelve months. . . .

Project Delta continues to be a particularly useful intelligence gathering means. . . . The technique of employment of these reconnaissance teams today differs somewhat in that they are now employed in VC-controlled areas where there are no major friendly actions in progress which might compromise Delta's presence. . . Currently these six-man US-Vietnamese teams "work an area" for longer periods (generally thirty days) than heretofore. Repeatedly they have collected valuable intelligence. . . .

Probably the single greatest U.S. shortcoming in Vietnam is our lack of timely, accurate intelligence. Soldiers' complaints about their repeated "walks in the woods" without contact give evidence of this problem. SF CIDG camps however, have helped to fill this vacuum. SF camps are able to establish effective agent nets in the locale of the camp using CIDG who are native to the area. . . . It is a unique capability which accrues to the USASF-CIDG system. MACV J2 states that over 50 percent of all their ground intelligence reports in the country come from Special Forces sources. . . The 1st Air Cavalry Division's Operation Crazy Horse during May 1966 in Binh Dinh Province is an example of an operation launched solely as a result of intelligence obtained by a CIDG-Special Forces patrol . . . Another example of success achieved from rapid exploitation of SF battlefield intelligence occurred when Camp Buon Ea Yang, Darlac Province, II CTZ, conducted an operation on 18 March 1966, in which a VC Company Commander was KIA and several documents were captured. Subsequent analysis of the documents indicated the location of four VC companies, approximately twenty-two kilometers from the camp. An operation was planned on the basis of the captured information. Operation Le Hai 21 made contact with a VC battalion located at the coordinates taken from the documents. . . .

The areas of civic action and psychological operations continue to occupy much of our attention. According to USARV records, the 5th Group Civic Action Program accounted for half of all the civic action projects conducted by USARV units during 1965-66 to date. . . .

In 1964 officials in the United States Mission in Saigon began pressing for more emphasis on providing advice and assistance to the Vietnamese on civil matters. One program which was developed and rapidly expanded called for U.S. advisers to assist the Vietnamese government officials in improving the civic and community activities in their local areas. The advice and assistance ranged from large-scale projects such as dam construction, crop development, bridge-building, and road improvements to the digging of wells, planning and supervising elementary sanitation systems, the establishment of small businesses, the construction of Montagnard hospitals, and the technical training of medical orderlies, dental technicians, and automotive mechanics.

U.S. Military Assistance Command, Vietnam, advisers were assigned to province or sector officials (roughly equivalent to state officials in the United States), and as the number of American advisers increased, they were assigned to the next lower political level, the subsector (town or village) officials. At the same time, it was recognized that the real focal point for advice and assistance should be the officials of the town or village. The availability of U.S. advisers, civilian and military, did not always stretch to include these grass roots officials.

Since a number of CIDG camps were located near subsector headquarters, the U.S. Army Special Forces was asked in 1964 to study the practicality of assigning the additional mission of advising subsector officials to the U.S. Special Forces commanders in the nearby CIDG camps. At this point in time, the 103 advisers from the Military Assistance Command, Vietnam, had proven their worth in the political structure but had all the responsibility they could manage. Geographically they were stretched thin. The U.S. Special Forces headquarters agreed to take on the additional mission of advising the Vietnamese civilian officials with the clear understanding that the CIDG combat mission had priority. One compelling reason for accepting these posts in a select number of locations was that the Special Forces adviser had a built-in defensive capability in the form of the CIDG troops in nearby camps, a resource not available to the Military Assistance Command adviser, especially in insecure districts largely controlled by the Viet Cong.

After a successful test period with an A detachment in this dual role, the Military Assistance Command assigned the subsector mission to appropriately situated A detachments in all four corps tactical zones. In certain provinces in III and IV Corps, where most of the subsector officials were advised by Special Forces A detachment commanders, the control B detachments were assigned the next higher political level requirement—the sector advisory mission. The mission of an A detachment commander in this assignment was to advise and assist the subsector official or district chief in the training and employment of his Regional and Popular Forces troops. As a sector adviser, a B detachment commander had a similar mission in relation to the sector commander or province chief.

By October 1965 five B detachments had coequal missions and thirty-eight A detachments were assigned the subsector mission. The

SPECIAL FORCES SOLDIERS BETWEEN MISSIONS

number of detachments assigned these missions peaked in the first quarter of 1966 at seven B and forty-one A detachments and thereafter declined. At the end of June 1967 there were four B and twenty-three A detachments so assigned. On the whole, the performance in the subsector and sector advisory mission by the Special Forces was very good, but A detachments were clearly better motivated and more effective in carrying out the subsector mission when controlled by a B detachment charged with the sector advisory mission.

The combination of the coequal missions of B detachments and the similarly charged A detachments under them was most productive in the Mekong Delta, where the presence of U.S. troops was not a factor, and the contest, despite presence of the Vietnam Army, was for the most part between government paramilitary forces and local Viet Cong units. In these circumstances, the B detachment commanders were in a position to plan and co-ordinate the operations of all CIDG, Regional Forces, and Popular Forces units in a province by using an integrated intelligence system. Detachment B–41 at Moc Hoa in Kien Tuong Province, B–42 at Chau Doc in the same province, and B-43 at Cao Lanh in Kien Phong Province —all in the IV Corps—were able to operate most effectively in this way.

In the area of combat developments, the achievements of Project Delta must be mentioned. Project Delta was the first unit specifically trained to perform special operations. When it first became operational in December 1964, after a long period of training, it consisted of six reconnaissance teams of eight Vietnamese and two U.S. Special Forces men each, and a reaction force, the Vietnam Army's 91st Ranger Battalion (Airborne) consisting of three companies. By 1967 Project Delta had expanded to sixteen reconnaissance teams composed of four Vietnamese and two U.S. Special Forces members, eight roadrunner teams, and a reaction force of six companies. The pattern of operations consisted of infiltrating teams, normally by helicopter, at dusk or after dark into a Viet Cong-controlled area, without benefit of lights or ground reception party. At first the teams were limited to reconnaissance and were withdrawn if discovered. Subsequently, a decision was made to allow them to continue operations provided that contact with the enemy had been safely broken and to attack small targets that they could handle without help. Missions were assigned by Military Assistance Command, Vietnam, or the Vietnamese Joint General Staff and were based on recommendations from the commanding general of the Vietnamese Special Forces and the commanding officer of the 5th Special Forces Group (Airborne).

Besides Project Delta, other combat developments in this period included the creation of the Apache Force and the Eagle Scouts, which were eventually integrated into the combat reconnaissance patrol and mobile strike forces, which also had their beginnings at this time. The full effectiveness and capabilities of the combat reconnaissance patrol and mobile strike forces were not realized until after June 1966.

The Apache Force was conceived as a combined force of Special Forces men and indigenous troops with the primary mission of orienting an American battalion or larger size unit prior to its commitment to combat against Viet Cong or North Vietnamese Army forces. The orientation included terrain walks, map analyses, Viet Cong or North Vietnamese small unit tactics, a review of lessons learned to date on enemy weaknesses and common mistakes made by U.S. forces when first committed. Finally, the Apache Force usually accompanied the American unit into the field for the first several days of combat. Its secondary mission was to serve as a multipurpose reserve for the Special Forces CIDG program in order

SOLDIERS FIRING AN 81-MM. MORTAR

to extend intelligence capabilities, conduct offensive operations with Nung companies (which came to be heavily utilized in the mobile strike forces), reinforce threatened areas, or act as a reaction force for camps, outposts, or forward operations bases under attack. The Eagle Scouts, like the Apache Force, were capable of reconnaissance and combat and could be moved by helicopter as well. These two forces, along with others, represented stages in the evolution of an effective reconnaissance and reaction force—the mobile strike force.

The formation of combat reconnaissance platoons of thirty-four men each, one platoon to each camp, began during the first quarter of 1965. It took some time to send all the platoons to Dong Ba Thin to receive special training at the Vietnamese Special Forces training center under Project Delta instructors, but the reconnaissance platoon became the elite unit of each camp and measurably increased the effectiveness of strike force operations. In most camps there was at least a reconnaissance squad attached to a regular strike force patrol, usually of company strength. The combat reconnaissance platoon was infrequently employed as a unit, but elements were often assigned the task of finding and fixing enemy targets or used for psychological operations and small raids, or to adjust artillery fire and air strikes. In July 1966 the decision was made to increase the strength to two reconnaissance platoons per CIDG camp.

From 1961 to 1964 a serious weakness in the CIDG program was the lack of troops to reinforce a camp garrison under attack or to exploit a patrol contact with the enemy.

General William C. Westmoreland in June 1965 approved the creation of a small reserve force for each C detachment for use in long-range patrolling, reinforcement, and reaction. These multipurpose reaction forces, called mobile strike forces, were formed during the fourth quarter of 1965. Another mobile strike force was organized at Nha Trang under the operational control of the Commanding Officer, 5th Special Forces Group (Airborne). Each mobile strike force consisted of a headquarters and three companies with a total strength of 594. Each company was composed of three infantry platoons, a weapons platoon, and a reconnaissance platoon which together had a total strength of 198. Mobile strike force troops were trained to a tactical competence beyond that of a CIDG camp strike force company. A Special Forces A detachment, initially without Vietnamese Special Forces counterpart, was assigned to each mobile strike force. At first full use was not made of the strike forces' capabilities as reaction and reconnaissance units. Instead these units for the most part acted as interior guards for the corps C detachments and the headquarters compound at Nha Trang. After June 1966, however, full use of the mobile strike forces was made, with excellent results.

There was a significant engagement at the A Shau camp in I Corps on 11 March 1966 that underscored the need for an effective reaction force. A Shau was an isolated camp southwest of Hue, about five miles from the Laotian border. The camp's mission was border surveillance and the interdiction of infiltration routes. During the first week in March 1966, captured enemy documents and information from defectors indicated that an attack by four North Vietnamese Army battalions was imminent. Reinforcements were requested. Headquarters, I Corps, disapproved the request, but the commander of the 5th Special Forces Group (Airborne) committed a mobile strike force company of 143 men, seven of them U.S. Special Forces men, which arrived in the camp on 7 March. The garrison consisted of 220 civilian irregulars, 10 U.S. Special Forces men, 6 Vietnamese Special Forces men, several interpreters, and 41 civilians.

Although local patrols and night ambush parties had failed to find the enemy on 7 and 8 March, the Viet Cong began to probe the outer defenses at 1930 on the 8th. Early on 9 March the enemy opened up with 81-mm. mortars, causing fifty-seven casualties and damaging camp structures. Air strikes were ineffective because of heavy ground fog, and one incoming C-47 crashed after being hit by ground fire. A helicopter evacuated twenty-six wounded.

Mortar and 57-mm. recoilless rifle fire beginning at 0400 on 10 March reduced most of the remaining buildings to rubble and silenced half the crew-served weapons. At 0500 the enemy launched heavy assaults across the runway against the east wall and—under cover of the tall grass that had been allowed to grow up in the defensive minefield—against the south wall. The defense of the southeast corner collapsed and the fighting surged into the camp. Survivors from the east wall and south wall defenses withdrew to positions near the communications bunker and the north wall at about 0830. Air strikes were then brought in with good effect on the enemy forces occupying part of the breached walls and on enemy units that were forming east of the airstrip for another assault, but they had no power to change the outcome of the fight. The enemy attack had succeeded. In midafternoon Headquarters, III Marine Amphibious Force, dispatched sixteen H-34 helicopters with tactical air support to evacuate the garrison. At 1720 survivors began to move toward the landing zone, with Special Forces mobile strike force troops fighting a rear guard action. Heavy enemy fire at the pickup point inflicted many casualities, and the waiting civilian irregulars panicked and tried to force their way into the aircraft. Two helicopters were destroyed by enemy fire. Some were unable to touch down because of the low ceiling. Only sixty-five of the original three to four hundred persons were evacuated. By 1745 all who remained and could do so—seven of the Special Forces men, forty mobile strike troops, fifty CIDG troops, and the two downed helicopter crews—made their escape, moving in a northeasterly direction. On 11 and 12 March several small groups were sighted by rescue aircraft and picked up. Further air search from the 13th through the 15th of March failed to locate any more survivors.

In July 1965 fighting took place near Camp Bong Son, Binh Dinh Province, II Corps, involving four Special Forces men and the 883d Vietnamese Regional Forces Company. The action bears mention because it illustrates the demands often made on Special Forces men in South Vietnam and reflects the considerable ability demonstrated by many Green Berets in combat. The following account was written by Major Paris D. Davis, then a captain, who participated in the action.

We had just finished a successful raid on a Viet Cong Regimental Headquarters, killing upwards of one hundred of the enemy. The raid had started shortly after midnight. We had four Americans and the 883rd Vietnamese Regional Force Company participating in the raid. After the raid was completed, the first platoon of the 883rd company broke and started to run just about the same time I gave the signal to pull in the security guarding the river bank. I went after the lead platoon, MSG Billy Waugh was with the second platoon, SSG David Morgan was with the third platoon, and SP4 Brown was with the fourth platoon. It was just beginning to get light (dawn) when I caught up to the first platoon and got them organized, and we were hit by automatic machine gun fire. It was up front and the main body of the platoon was hit by the machine gun. I was hit in the hand by a fragment from a hand grenade. About the time I started moving the platoon back to the main body, I heard firing and saw a wounded friendly VN soldier running from the direction of the firing. He told me that the remainder of the 883rd company was under attack. I moved the platoon I had back towards the main body. When I reached the company, the enemy had it pinned down in an open field with automatic weapons and mortar fire.

I immediately ordered the platoon I had to return the fire, but they did not—only a few men fired. I started firing at the enemy, moving up and down the line, encouraging the 883rd company to return the fire. We started to receive fire from the right flank. I ran down to where the firing was and found five Viet Cong coming over the trench line. I killed all five, and then I heard firing from the left flank. I ran down there and saw about six Viet Cong moving toward our position. I threw a grenade and killed four of them. My M16 jammed, so I shot one with my pistol and hit the other with my M16 again and again until he was dead.

MSG Waugh started to yell that he had been shot in the foot. I ran to the middle of the open field and tried to get MSG Waugh, but the Viet Cong automatic fire was too intense, and I had to move back to safety. By this time SSG Morgan, who was at the edge of the open field, came to. He had been knocked out by a VC mortar round. He told me that he was receiving sniper fire. I spotted the sniper, and shot him in his camouflaged man-hole. I crawled over and dropped a grenade in the hole killing two additional Viet Cong.

I was able at this time to make contact with the FAC, CPT Bronson and SGT Ronald Dies. CPT Bronson diverted a flight of 105's and had them drop their bombs on the enemy's position. I ran out and pulled SSG Morgan to safety. He was slightly wounded, and I treated him for shock. The enemy again tried to overrun our position. I picked up a machine gun and started firing. I saw four or five of the enemy drop and the remaining ones break and run. I then set up the 60mm mortar, dropped about five or six mortars down the tube, and ran out and tried to get MSG Waugh. SSG Morgan was partially recovered and placing machine gun fire into the enemy position. I ran out and tried to pick up MSG Waugh, who had by now been wounded four times in his right foot. I tried to pick him up, but I was unable to do so. I was shot slightly in the back of my leg as I ran for cover. By this time CPT Bronson had gotten a flight of F4's. They started to drop bombs on the enemy. I ran out again, and this time was shot in the wrist—but I was able to pick up MSG Waugh and carried him fireman style, in a hail of automatic weapon fire, to safety. I called for a MEDEVAC for MSG Waugh. When the MEDEVAC came, I carried MSG Waugh about 200 yards up over a hill. As I put MSG Waugh on the helicopter, SFC Reinburg got off the ship and ran down to where the 883rd company was located. He was shot through the chest almost immediately. I ran to where he was and gave him first aid. With SSG Morgan's help, I pulled him to safety.

The enemy again tried to overrun our position. I picked up the nearest weapon and started to fire. I was also throwing grenades. I killed about six or seven. I was then ordered to take the troops I had and leave. I informed the colonel in the C&C ship that I had one wounded American and one American I didn't know the status of. I informed the colonel that I would not leave until I got all the Americans out. SFC Reinburg was MEDEVACed out. The fighting continued until mid-afternoon. We could not get the company we had to fight. The enemy tried to overrun our position two more times. We finally got reinforcements, and with them I was able to go out and get SP4 Brown who lay out in the middle of the field some fourteen hours from the start until the close of the battle.

Major Davis received the Silver Star and the Purple Heart for his efforts in this action.

CHAPTER V

The Combined Special Forces CIDG On the Offensive

Between May 1966 and May 1967 there were wide-ranging changes and improvements in the operational employment of the Special Forces and the Civilian Irregular Defense Group program assets. Areas in which substantial progress was made during these twelve months were enumerated in the debriefing report of 31 May 1967.

Emphasis was placed on mobility in operations in all camps, the report notes. Mobile guerrilla forces were formed and operated in enemy-controlled zones; mobile strike forces doubled in number, were airborne-qualified, and participated in many operations as exploitation forces and as reserves for camps needing additional strength at critical times. Two or more CIDG camps conducted operations within the same corps tactical zone or across corps boundaries.

The CIDG program grew. Twenty-two new camps opened and nine camps in relatively pacified areas were closed; the new camps were constructed as "fighting camps" and, in the case of the delta region, "floating camps." The number of CIDG combat reconnaissance platoons was increased from thirty-four to seventy-three. The Military Assistance Command Recondo School to prepare selected soldiers for long-range reconnaissance duty was established in September 1966 with sixty students per three-week course; by January 1967 the number of students had doubled.

Operational responsibility for Camp Plei Mrong was transferred to exclusive Vietnamese Special Forces control, with supervision by Detachment B–24 (Kontum) and Company B (Pleiku). This was the first time a CIDG camp had been turned over to the Vietnamese Special Forces and may be viewed as a first step in so-called Vietnamization. Operations were integrated with nearby U.S. and Free World forces with surprisingly good results. The integration was most evident in operations and intelligence and pointed up the need for further effort in the communications field.

A combined U.S. Special Forces–Vietnamese Special Forces operational directive which specified that camp operations begin and, where possible, end in the hours of darkness was complied with by the CIDG camps. This tactic, based on the realization that the night belongs to him who uses it, was a sharp and welcome departure from previous practice. The substantial jump in the number of the enemy killed in the last quarter of 1966 (817 to 1,302) was unquestionably influenced by this change in tactics.

Procedures were developed to produce annually a U.S. Special Forces campaign plan to co-ordinate future camp installations. A campaign plan for fiscal year 1967 was produced in September 1966, and one for fiscal year 1968 was completed in July 1967. This development grew out of a command conference in Nha Trang in August 1966 at which General Westmoreland directed the commander of the 5th Special Forces Group to make a close examination of the present and proposed deployment of the group's operational detachments throughout Vietnam. He specified that each detachment be examined to insure that it had a mission and a location that would enable it to exert its full potential. He suggested that A detachments and their civilian irregular strike forces be replaced where practicable by South Vietnamese Army or Regional Forces and Popular Forces units and that any CIDG camp improperly located to carry out its mission be relocated. Planning was to be co-ordinated with corps senior advisers and Vietnamese corps commanders.

A flood campaign plan for operations in the delta region during flood conditions was developed. The plan was built around use of some four hundred watercraft (including eighty-four airboats), helicopters, sophisticated U.S. Navy craft, and waterborne maneuvers, with civilian irregular forces in the boats.

Projects Omega and Sigma were formed and, along with Project Delta, conducted operations in the field.

The 5th Special Forces Group administrative and logistic organization was critically examined with the object of improving control and supervision. A number of elements were created and installed as new parts of the staff including new comptroller, judge advocate, aviation, Air Force liaison, and staff engineer sections, and a new acting inspector general. The existing S–2 (intelligence) section was completely overhauled, and the S–3 section as well as the open mess association was revised. New radio research and historical units were created, as was a new military intelligence augmentation detachment.

New funding controls were imposed and inspected, and new

athletic, morale, and welfare facilities, including clubs, a theater, a chapel, and a library were proposed. The promotion system for enlisted men was revised. Noncommissioned officers conducted all group briefings, including briefings for the Chief of Staff, U.S. Army. Officers in staff positions were replaced whenever possible by noncommissioned officers, with the greatest success in the area specialist (S–3) and order of battle (S–2) specialties.

Psychological operations were withdrawn from the S–5 section and installed in the S–3 section, where they could be integrated with all operations. The S–5 section was abolished in favor of a revolutionary development support activities section.

Liaison officers were exchanged with the Vietnamese Special Forces high command, and U.S. Special Forces liaison officers were attached to all major U.S. Army units close to CIDG camps.

Other changes were made in decentralizing logistics, including the creation of forward supply points in all four corps tactical zones, plus direct sea, air, and road shipments thereto. A detailed civic action program was developed in an attempt to avoid philosophy and show "how to do it" instead. Handbooks on how to operate any portion of the group's business from an A detachment to an agent net were made up and distributed. A new training program for civilian irregulars and new techniques for resupply of CIDG camps weathered out or under violent attack were developed.

Problems noted in the debriefing report of May 1967 included the adverse effects produced by the continuing prejudice on the part of U.S. forces against the Vietnam Army and the continuing danger of Montagnard rebellion.

The campaign plan to co-ordinate future camp installations provided for the use of assigned U.S. Special Forces troops by positioning them for optimum employment and co-ordinated action and by giving them tasks commensurate with their capabilities.

The concept of operations plan for 1967 included the following tenets: offensive operations were to be the principal means to achieve over-all objectives. Emphasis would be placed on the Revolutionary Development Support Program—designed to win the people to the government—assisting Free World forces, and increasing the participation of the Vietnam Army. The plan, presented in detail for each corps tactical zone, projected the conversion of seventeen CIDG camps to the Regional Forces. Efforts were called for in three major areas: strategic deployment of CIDG camps in the four corps tactical zones; employment of mobile guerrilla, Special Forces, and CIDG troops in areas not covered by CIDG camp operations; and use of long-range reconnaissance forces provided to the Vietnamese Army and Free World forces, including training to develop their organic reconnaissance capabilities.

Intelligence Operations Overhauled

In the course of a year, beginning in June 1966, the intelligence project of the 5th Special Forces Group (Airborne) evolved from a fragmented effort into a co-ordinated, productive program on a countrywide scale. First, emphasis was placed upon establishing American–Vietnamese Special Forces intelligence projects at all A and B detachments. These served to improve the intelligence skills of the Vietnamese Special Forces and began to prepare them for the time when they would be on their own. One problem which had to be met was that the quantity and quality of information coming out of the A detachments varied widely. The flow of information was crucial to the success of the intelligence effort and required improvement. Probably the most serious problem that plagued the Special Forces intelligence program was the inability of the 5th Group to analyze rapidly and disseminate quickly the information that came in so that combat forces could act promptly on known data about the enemy. Also the sheer volume of intelligence information which began to flow into the group from the various agencies that accompanied the U.S. buildup required a new organizational structure.

To set the intelligence house in order, at group headquarters a number of regulations were written that specified exactly the steps to be taken. Regulations were written on source control, collection procedures, intelligence reporting, methods of recording, and dissemination. Within the S–2 section, a counterintelligence branch and a collection branch were created in addition to the order of battle branch and administrative branch. After distribution of the new regulations, mobile training teams were sent into the field, first to the A detachments, to assist in carrying out the new procedures. The whole intelligence program at an A detachment was usually embodied in one individual, the intelligence sergeant, whose training and background varied. The intelligence sergeant normally acquired a large store of valuable information about the enemy in his detachment's tactical area of responsibility, but often he took this information with him when he left the detachment, forcing his replacement to start from scratch. The new regulations and mobile training teams provided instructions for developing agent nets to gather information on enemy operating bases and underground organization and required the A and B detachments to maintain handbooks, overlays, and organizational charts that would provide permanent, continuing files of verified information about the enemy. As the system of gathering, classifying, and recording information on the enemy began to function at the A and B detachment levels, operational results began to show. It became the practice for camps to plan operations based on knowledge of where the enemy was to be found instead of trying to conduct operations to cover the entire tactical area of responsibility of the camp.

To exploit the information that was being gathered by the A and B detachments, it was necessary to refine the process of analyzing the raw information in order to produce finished intelligence and then to distribute it to lower, higher, and adjacent headquarters. This task was accomplished by the establishment of intelligence analysis centers, which operated under the S–2 at each of the four Special Forces company headquarters located in the four corps tactical zones. Each center had four branches: recording, analysis and evaluation, collection, and dissemination. The major advantage gained from the intelligence analysis centers was that the analysis and dissemination functions were decentralized and therefore became much more responsive and efficient.

Up to this point, the intelligence program had been put into effect by men from within the 5th Special Forces Group. More people were needed, however, and these were obtained by adding to the 5th Group the 403d Radio Research Special Operations Detachment and an unnumbered 110-man military intelligence detachment. The members of the military intelligence detachment were subdivided into five composite teams, each containing men with counterintelligence, interrogation, collection, analysis, and administrative skills. One team was kept at group headquarters and one team went to each of the four company headquarters in the four corps tactical zones. Working in conjunction with other Special Forces troops, most of the collection and counterintelligence men were attached to the subordinate A and B detachments by the companies. At these locations they would assist in writing collection plans; in recruiting, training, and assigning operations to agents; and in developing counterintelligence and counterespionage nets. The nets were especially important for identifying and eliminating enemy agents who had penetrated the indigenous forces in the CIDG camp and were a substantial threat to camp security. The group headquarters, with its augmentation team, directed these programs and provided policy guidance at all levels of the command.

With the deployment of the military intelligence augmentation teams, the group could finally operate effectively with other U.S. intelligence organizations. It became possible to accept the teams' information, collate it with the information from Special Forces channels, and disseminate a higher quality of intelligence through the company intelligence analysis centers to subordinate B and A detachments, as well as to U.S. and other Free World Military Assistance Forces operating in each of the company areas of interest.

The status of the Vietnamese Special Forces in the eyes of many U.S. Special Forces men had been up to this point rather low, and not without justification. Nevertheless, the Vietnamese Special Forces had been improving and the results began to show. The Vietnamese Special Forces organization consisted of a Special Forces command, which was composed of a headquarters element, one Special Forces group, an airborne Ranger battalion, a Special Forces training center, a signal company, a headquarters and service company, and Project Delta. The Vietnamese Special Forces Group was organized into four C detachments—one to each corps tactical zone—twelve B detachments, and seventy-three A detachments. The Vietnamese Special Forces was charged by the Joint General Staff with the following missions: to plan and conduct unconventional warfare operations when approved by the Joint General Staff; to plan, conduct, and support counterinsurgency operations, such as the CIDG program, within the republic; to collect, process, and submit to the Joint General Staff intelligence information; to carry out political warfare activities; and, for the C detachments, to advise the commander of a corps tactical zone on the employment of Special Forces and civilian irregular troops.

A marked increase in the competence of the Vietnamese Special Forces, with the most noticeable improvements at the A detachment level, became apparent during 1966 and 1967. It finally became possible to begin the long-desired gradual reduction in the 5th Special Forces Group advisory participation in the CIDG program. For many years Special Forces detachment commanders, although officially only advisers, actually served as camp commanders because of the lack of leadership ability on the part of their Vietnamese counterparts. During 1967 the importance of the advisory aspect of the Special Forces mission was emphasized at the command level in order to give the Vietnamese Special Forces A team commander more responsibility. Results were generally positive, with Vietnamese Special Forces A detachment commanders assuming greater responsibility and demonstrating improved leadership. The single best example of the new Vietnamese leadership at work was Camp

Plei Mrong. On 1 May 1967 the camp was turned over completely to the control of the Vietnamese Special Forces, and the U.S. Special Forces detachment was withdrawn. Further, plans were made for the turnover of two more camps, Minh Thanh and Vinh Gia, during the last part of 1967 depending on how well things went at Plei Mrong. Vinh Gia was successfully transferred to Vietnamese Special Forces control on 27 June 1967, and Minh Thanh shortly thereafter.

The typical CIDG camp in 1967 was commanded by a Vietnamese Special Forces A team commander. The organization of his A team paralleled that of the U.S. Special Forces A team, with men trained in operations, intelligence, medical aid, weapons, political warfare, communications, supply, and demolitions. Each of the Vietnamese Special Forces team members in a camp was advised by a U.S. Special Forces team member with similar skills. The normal force contingent at a CIDG camp, besides the Special Forces troops, consisted of four 132-man CIDG companies, two combat reconnaissance platoons, a civic action and psychological operations squad; and could include a recoilless rifle or a 105-mm. artillery section.

A countrywide plan for the deployment of CIDG assets was included in the 1967 concept of operations for the 5th Special Forces Group (Airborne) and approved by General Westmoreland on 7 March 1967. (*Map 6*)

In the I Corps Tactical Zone the plan was to maintain and expand the present camps to the west of the Free World Military Assistance Forces. In II Corps it called for maintaining and reinforcing the system of border surveillance camps and building a second line of camps in the interior highlands. The primary purpose of this second line of camps was to interdict the infiltration routes leading to and from the populated coastal region. In III Corps the plan was to maintain and reinforce camps engaged in border surveillance and to place camps in, near, and between War Zones C and D in order to better execute the missions of interdiction of infiltration routes and harassment of base areas. In IV Corps the plan was to maintain and reinforce the interdiction of Viet Cong infiltration routes from the manpower and material reservoir of the Mekong Delta and to clear the Plain of Reeds of Viet Cong known to be operating there.

Except in II Corps, this plan of action was successfully carried out in 1966 and 1967. Construction was begun in I Corps of one of the two camps approved by General Westmoreland. In III Corps five of the proposed nine camps were begun, with the remainder waiting for engineers to become available. Of the five camps sched-

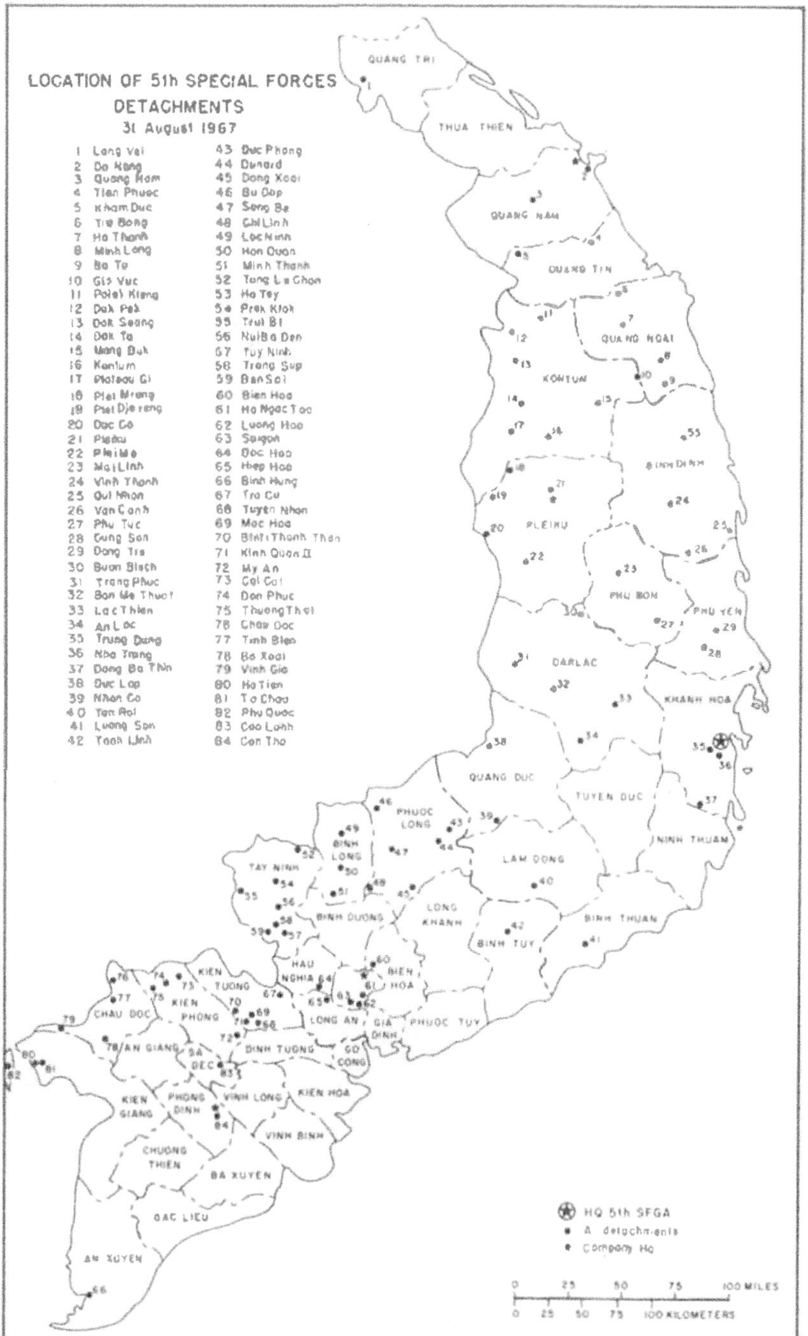

LOCATION OF 5th SPECIAL FORCES
DETACHMENTS
31 August 1967
1 Lang Vei
2 Da Nang
3 Quang Nam
4 Tien Phuoc
5 Kham Duc
6 Tra Bong
7 Ha Thanh
8 Minh Long
9 Ba To
10 Gia Vuc
11 Polei Kleng
12 Dak Pek
13 Dak Seang
14 Dak To
15 Mang Buk
16 Kontum
17 Plateau Gi
18 Plei Mrong
19 Plei Djereng
20 Duc Co
21 Pleiku
22 Plei Me
23 Mai Linh
24 Vinh Thanh
25 Qui Nhon
26 Van Canh
27 Phu Tuc
28 Cung Son
29 Dong Tre
30 Buon Blech
31 Trang Phuc
32 Ban Me Thuot
33 Lac Thien
34 An Loc
35 Trung Dung
36 Nha Trang
37 Dong Ba Thin
38 Duc Lap
39 Nhon Co
40 Tan Rai
41 Luong Son
42 Tanh Linh
43 Duc Phong
44 Dunard
45 Dong Xoai
46 Bu Dop
47 Song Be
48 Chi Linh
49 Loc Ninh
50 Hon Quan
51 Minh Thanh
52 Tong Le Chon
53 Ha Tay
54 Prek Klok
55 Trai Bi
56 Nui Ba Den
57 Tuy Ninh
58 Trang Sup
59 Ben Soi
60 Bien Hoa
61 Ho Ngoc Tao
62 Luong Hoa
63 Saigon
64 Duc Hoa
65 Hiep Hoa
66 Binh Hung
67 Tra Cu
68 Tuyen Nhon
69 Moc Hoa
70 Binh Thanh Thon
71 Kinh Quan II
72 My An
73 Cai Cai
74 Don Phuc
75 Thuong Thoi
76 Chau Doc
77 Tinh Bien
78 Ba Xoai
79 Vinh Gia
80 Ha Tien
81 To Chau
82 Phu Quoc
83 Cao Lanh
84 Can Tho
QUANG TRI
THUA THIEN
QUANG NAM
QUANG TIN
QUANG NGAI
KONTUM
BINH DINH
PLEIKU
PHU BON
PHU YEN
DARLAC
KHANH HOA
QUANG DUC
TUYEN DUC
NINH THUAN
PHUOC LONG
LAM DONG
BINH LONG
TAY NINH
BINH DUONG
LONG KHANH
BINH TUY
BINH THUAN
HAU NGHIA
BIEN HOA
KIEN TUONG
KIEN PHONG
CHAU DOC
LONG AN
GIA DINH
PHUOC TUY
AN GIANG
SA DEC
DINH TUONG
GO CONG
KIEN GIANG
PHONG DINH
VINH LONG
KIEN HOA
VINH BINH
CHUONG THIEN
BA XUYEN
BAC LIEU
AN XUYEN
HQ 5th SFGA
A detachments
Company HQ
0 25 50 75 100 MILES
0 25 50 75 100 KILOMETERS

MAP 6

uled to be closed or converted, only one remained in operation by the summer of 1967. In IV Corps by the summer of 1967 three of the seven proposed camps were open and conducting operations. While the turnover went according to plan in II Corps at Plei Mrong, the construction of new camps in the interior did not take place because of the changes in the deployment of Free World Military Assistance Forces.

The quality of the CIDG troops improved during this period also. The debriefing notes of the 5th Special Forces Group commander attested to the improvement.

> During my tour in Vietnam, I have seen a steady improvement in the fighting spirit and military proficiency of the CIDG. This is directly attributable to the improvements previously noted in the VNSF A detachments. Increased tactical success and night operations have given the CIDG greater confidence. Even in cases when USASF and VNSF personnel are not present, the CIDG are closing with and destroying enemy units. The most notable example of this occurred on 8 April 1967. Three platoons from Camp Trung Dung (A-502), without US or VNSF present, were deployed south and east of the camp in night ambush position, when an NVA company attacked an RF outpost. Employing superb fire and maneuver, and coordinating with each other by radio, the CIDG platoons moved to cut off the enemy withdrawal, killing forty-two NVA and capturing twenty weapons. While this is only one incident, it is indicative of the improvement in the CIDG forces.

The tactical and operational innovations of this period improved significantly the performance of the civilian irregulars. To determine progress in a counterinsurgency is difficult at best. There are many intangibles that cannot be reflected in statistics, but statistics can be used to determine whether progress was made in a given area. (*Table 7*) The increase in combat effectiveness was a

TABLE 7—COMPARATIVE DATA ON COMBAT OPERATIONS, 1966–1967

Item	1st Quarter 1966	1st Quarter 1967	Percentage of Change
All operations	72,597	78,249	+7
Night operations	24,920	29,010	+16
Contacts	878	1,061	+21
Enemy killed (body count)	1,348	1,912	+42
Enemy captured	595	445	−25
Weapons captured	431	585	+37
USASF/CIDG killed	616	446	−28
USASF/CIDG wounded	344	1,080	+215
USASF/CIDG missing	319	319	..
Weapons lost	432	344	−20

direct result of the tactical and operational command directives initiated by the 5th Special Forces Group (Airborne).

While a strong effort was made to increase the number of operations, command emphasis centered on improving techniques in three major respects. First, it was desirable to increase operations based on solid intelligence about the enemy rather than random search methods. While such a tactic may seem basic, it had not been properly used. By the new procedure Headquarters, 5th Special Forces Group (Airborne), retained a plot of all enemy sightings within each camp's area of responsibility. Superimposed over this plot was a second plot of all company search and destroy operations conducted by the camp. At the end of each month an analysis of these plots soon disclosed whether the camps were, in fact, operating properly. Second, an effort was made to conduct operations over a greater range of territory and to sustain them longer. Third, both the Vietnamese Special Forces high command and 5th Special Forces Group headquarters directed that in all cases operations were to begin and, where feasible, end during the hours of darkness. The only exceptions to this policy were those heliborne, reaction, and other special operations that had to be initiated during periods of daylight. By the spring of 1967 well over 90 percent of the operations conducted by units subordinate to, and advised by, the 5th Group were begun during the hours of darkness. The effect of this program was a marked increase in the number of enemy contacts developed by the CIDG, a decrease in the number of enemy attacks on installations, and the mounting confidence that the irregulars and Vietnamese Special Forces had in their own ability to operate at night. As success and confidence grew, so did the effectiveness of night operations, with the result that the Viet Cong lost their most valuable protection—the cloak of darkness.

The increase in the number of the enemy killed and the comparative decrease in CIDG and Special Forces losses were directly attributable to the improved techniques. By carrying the fight to the enemy instead of waiting for him to pick the situation, the Special Forces and civilian irregulars were able to employ their supporting fires effectively against an unprepared opponent.

The decided increase in the number of civilian irregulars wounded was a direct result of the new procedure of conducting more operations in enemy base areas that were extensively booby-trapped. Soldiers moving into and through those areas were bound to take more casualties. Fortunately, effective medical evacuation and swift treatment for the wounded cut these losses. The increase

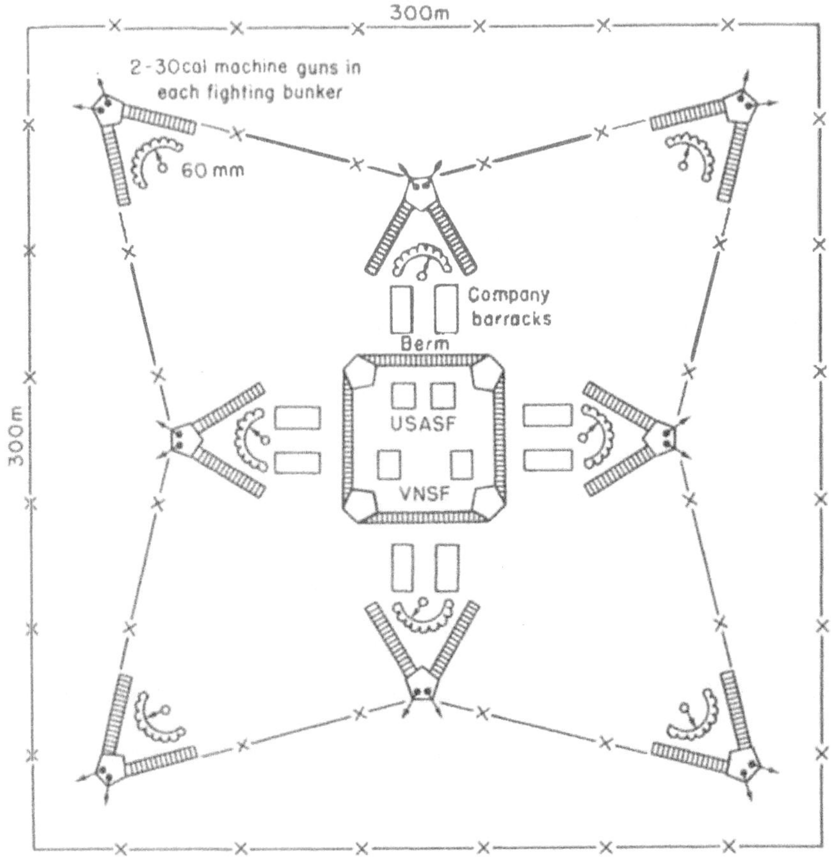

DIAGRAM OF A FIGHTING CAMP

in the number of weapons captured and the corresponding decrease in the number of weapons lost is an indication that CIDG troops were standing and fighting, while the enemy was withdrawing from the battlefield and leaving his dead and weapons behind.

The physical facilities and configuration of the camps underwent a significant change in the course of the year with the development of first the fighting camp and then the floating camp. The fighting camp was an austere, functional, easily defended camp, which employed the principle of defense in depth. It was designed not as an isolated, walled fortress in the midst of hostile territory, but as a base for extended operations throughout the A detachment tactical area of responsibility. The fighting camp concept was based upon aggressive combat operations backed up by a capability to

VIETNAMESE STRIKE TROOPS MAN AIRBOAT IN RIVER OPERATION

reinforce rapidly. Both kinds of camps were constructed with locally procured materials and labor at a saving of approximately ten thousand dollars per camp over the previous camp construction costs. The first camp to be built according to fighting camp specifications was at Plei Djereng, A–251, Pleiku Province, in II Corps Tactical Zone. Construction was begun on 13 December 1966 by the Special Forces Engineer Team KB–7 of the 539th Engineer Detachment, engineer augmentation for the 5th Special Forces Group. The floating camps were constructed in areas inundated annually by the floodwaters of the Mekong River. To insure that the camps would be able to function in spite of the flooding, buildings were constructed a story and a half high, with a floating floor that rose with the water; medical bunkers, ammunition bunkers, and crew-served weapons positions were built on reinforced platforms that also floated; and floating helipads were made capable of supporting a loaded UH–1D.

In order to maintain operational efficiency during those periods when troop changes were taking place at remote Special Forces camps, and to pass on all the accumulated knowledge and experience of past Special Forces operations to new camp commanders, two A detachment handbooks were published. The first book furnished guidelines for the opening of new CIDG camps, the second established procedures for the operation of those camps. The handbooks were distributed to every level of command in the 5th Special Forces Group. They proved to be an invaluable aid for new A detachment commanders in strengthening the security and improving the operational effectiveness of their camps.

In the fall of 1966, before floating camps were built, floods in IV Corps threatened to destroy some camps and curtail operations throughout the area. By extending the Special Forces logistical capability to its maximum and by improvising such things as floating helicopter pads and weapons emplacements, camps were kept in operation even though some of them were completely submerged. A completely new set of tactics was developed employing the speed and firepower of airboats, U.S. Navy river patrol boats, patrol air cushion vehicles, armed helicopters, and reconnaissance aircraft in combined operations supported by artillery and tactical aircraft. The results were telling victories over the Viet Cong and proof that with imagination and perseverance the flood season could be turned into a significant tactical advantage.

In terms of troop strength, the CIDG program continued to grow in this year, also. The total number of companies increased from 249 to 268. Combat reconnaissance platoons, thirty-four civilian irregulars per platoon, were authorized: these were to be increased from one to two platoons for each A detachment. Twelve 106-mm. recoilless rifle sections of six or eight men each were added, as were twenty-two howitzers with fourteen-man crews.

Probably the most significant achievement made in the whole CIDG program was the development of strategic concepts for the deployment of CIDG camps. These concepts were put into effect with the closing or conversion of camps that were not needed, and the establishment of new camps which supported the strategic plan in each corps tactical zone. In the course of the year, twenty-two new camps were opened in and around enemy base areas, while nine camps were closed or converted to Regional Forces or Popular Forces status.

Substantial gains were made in the effectiveness of the Special Forces program during 1966 and 1967. The war finally was carried to the enemy, into his base areas, at Special Forces initiative. In May 1966 over 50 percent of CIDG camps had surrounding areas that could not be entered because of the strength of the enemy. By May 1967, because of the increased combat power of the Special

LAST-MINUTE BRIEFING OF STRIKE FORCE BEFORE ATTACK

Forces and civilian irregulars, friendly forces were capable of operating throughout areas of responsibility with minimal losses. This progress was not without cost, however; fifty-five Special Forces and 1,654 Vietnamese, including the civilian irregulars, gave their lives during the year. There were 7,000 of the enemy killed, according to body count, and many more who were not counted.

During February, March, and April of 1967, combat forces advised by the 5th Special Forces Group expanded and intensified their operations in remote areas previously under the control of the Viet Cong. Five new camps were opened in the midst of traditional enemy strongholds. In III Corps the establishment of Camps Prek Klok and Tong Le Chon, in War Zone C, was an important step in restricting the enemy's use of that notorious base area. In several cases the enemy strongly opposed these intrusions into what he had considered his territory. On 14 April 1967 Camp Prek Klok received approximately 150 rounds of 82-mm. mortar fire, followed by a ground attack by an estimated two battalions of Viet Cong. This was the first large-scale attack on a Special Forces camp since A Shau was overrun in March 1966. In the vicinity of Camps My Phuoc Tay and My An, there were numerous encounters with Viet Cong units of company and battalion size. Because of the aggressive tactics of the CIDG and mobile strike force units and the rapid reaction of tactical air and artillery support, the Viet Cong suffered several defeats in this area.

In I Corps on 4 May 1967 at 0330 Camp Lang Vei, Detachment A-101, Quang Tri Province, was attacked by a company-size force supported by mortars and tanks. About one platoon of Viet Cong gained entry into the camp. With the assistance of fire support from Khe Sanh, enemy elements were repelled from the camp at 0500. Two Special Forces men were killed and five wounded; seventeen civilian irregulars were killed, thirty-five wounded, and thirty-eight missing. Enemy losses were seven killed and five wounded. Subsequent intelligence and prisoner of war interrogations indicated that the attackers were aided from inside the camp by Viet Cong who had infiltrated the CIDG units, posing as recruits. One prisoner of war said that he had been contacted by the Viet Cong before the attack and directed to join the CIDG at Lang Vei in order to obtain information on the camp. After joining the CIDG, the man recruited four other civilian irregulars to assist him. One man was to determine the locations of all bunkers within the camp, the second was to report on all the guard positions and how well the posts were manned, the third was to make a sketch of the camp, and the fourth was to report on supplies brought into the camp from Khe Sanh. The Viet Cong had contacted the prisoner who was under questioning on four occasions before the 4 May attack to get the information. On the night of the attack, the prisoner of war and another CIDG man killed two of the camp guards and led the Viet Cong force through the wire and minefield defenses into the camp's perimeter. This technique of prior infiltration was a Viet Cong tactic common to almost every attack on a camp.

On 8 May 1967 Camp Con Thien, Detachment A-110, Quang Tri Province, was attacked at 0230 by two battalions of the 812th North Vietnam Army Regiment with sappers attached. The assault was supported by mortars, rockets, and flame throwers. Artillery and air support was employed against the enemy throughout the attack. Since the camp was important to, and forward of, the Marine Corps positions in the demilitarized zone area of I Corps, a Marine battalion was moved into the camp for defensive strength. The perimeter, which was defended jointly by the U.S. Marine Corps, U.S. Special Forces, Vietnamese Special Forces and civilian irregulars, was penetrated at two places. Of the 212 of the enemy killed in the attack, 38 were credited to Special Forces, Vietnamese Special Forces, and CIDG troops. The defenders lost 14 CIDG men killed, 4 Special Forces and 16 CIDG men wounded, and two CIDG men missing, along with 44 U.S. marines killed and 110 marines and 5 men from a U.S. Navy construction battalion wounded. Enemy weapons captured included 4 flame throwers, 4 crew-served weapons, 12 40-mm. rockets, and over 100 individual weapons. Throughout the month of May the camp was subjected to constant harassing fire by mortars, artillery, rockets, and recoilless rifles. The total number of rounds received was over 1,500; on one occasion some 250 rounds landed in the space of four minutes.

The following chronological summary of the action at Con Thien was written by the Special Forces A detachment commander, Captain Craig R. Chamberlain, and submitted with the monthly operational summary of Detachment A–110 for the month of May 1967:

0210 hrs. —Six to eight round burst of small arms fire was heard east of camp. No other sounds were heard. No particular significance was given to this fact at the time because the enemy had fired weapons around the position on previous occasions apparently to draw fire in order to better determine the automatic weapon positions.

0230 hrs. —Attack commenced: I was sitting in the entrance to the command bunker when mortar and recoilless rifle fire of great intensity started landing throughout the USSF position and along the perimeter. Although some minutes must have passed, it seemed that small arms fire broke out along the perimeter almost immediately.

0245 hrs. —SSG Gibson, radio supervisor, and Kiet, camp interpreter, joined me at the CP bunker. Throughout this time a constant monitor was made of the USMC Battalion Tactical net. Situation at this time was still not clear and scope of enemy attack not fully appreciated.

0255 hrs. —SSG Brillante and SGT Zicaro joined the group at the CP bunker. Enemy by this time had penetrated to Co 146 position and sappers were moving into USSF/LLDB (*Lac Luong Dac Biet*—Vietnamese name for the VNSF) area. Some confusion existed at this time because of the difficulty in determining if the moving personnel were NVA or CIDG, who having been pushed off the perimeter, were looking for a place to regroup. Within a few moments we were able to determine who the enemy were, and the NCO's at my position took them under fire with small arms and grenades.

0300-0320 hrs.—Fighting continued around the bunker with exchange of grenades and small arms fire. At least six homemade grenades exploded outside the CP bunker but did no damage. One AT round from a B-40 hit the north side of the bunker and penetrated through the sandbags into the connex box that formed the base for the bunker. Throughout this period we were able to hold the CP bunker and had suffered only minor wounds to USASF personnel.

0320 hrs. —Enemy moved a flame thrower to a position on the right front of the CP bunker, firing from about 10 to 15 meters range. They put the first shot right in front of the CP bunker, but for some reason ignition of the fuel was not complete. There was a flash and roar, and it was immediately decided to abandon the position.

0330 hrs. —Relocated CP group into the Sea Bee's area. Until our arrival the Sea Bee's had not realized the situation and were still deep in their bunkers. They were immediately informed of the situation and moved into the trenches, and a secondary strong point was created. At this time coordination and communication was reestablished with Camp Commander and the LLDB detachment. By using one of their radios, communication was reestablished on the battalion tactical net.

0335 hrs. —Remainder of the USASF Det (CPT King, SFC Loff, SFC Lansberry, SFC Gomez) joined us at the Sea Bee's position. They had been located on top of hill 158 in positions within the area of construction for the new camp. They too had been confused as to the situation, and it was not until they observed the flame thrower being used that they realized the seriousness of the situation. NOTE: (Illumination throughout the action was very sporadic. The supply of hand-held flares was rapidly exhausted, and the USASF 81mm mortar position was untenable for all practical purposes during this phase of the action. A flare ship was on station a good portion of the time but was having difficulty in getting oriented over the position. This, coupled with low ceiling and smoke, made its effectiveness very limited. The importance of illuminations were obvious. When there was sufficient light, definite lulls in the fighting, particularly along the perimeter, would occur, only to be followed by intense action as soon as the flares burned out.)

0400-0500 hrs.—USASF and Sea Bees remained in position, and we satisfied ourselves with keeping the enemy out of our position. Fighting continued to be very heavy along the Marine perimeter. Some difficulty was encountered in keeping abreast of the USMC situation. During this period, ammo was resupplied to them using amtracks. The Bn Commander utilized the Marine Engineers to reinforce and strengthen their position. Some confusion and shooting took place as the amtracks passed in front of the CIDG positions. No confusion existed regarding whose tracks they were, but rather in regards to the ground element moving with them. The camp commander was able to get things under control very rapidly once he was made aware of the situation. NOTE: (During the period there were undoubtedly times when we could have brought enemy forces under fire from our position, but due to the confusion, I directed all USASF and Sea Bee personnel to hold their fire until I was absolutely sure of the target. This occurred mainly in the Marine position in the east, just north of where the airstrip bisected the perimeter.)

0500-0630 hrs.—The situation had clarified itself enough at this time so that we were able to take steps to push the enemy elements out of our immediate area. During this time, the .50 cal MG position was reoccupied and effective fire placed on the enemy troops occupying part of the Marine perimeter as well as a portion of the CIDG perimeter. This weapon kept going until mechanical failure, coupled with personnel being wounded, forced us to leave the gun and return to a secure position where we were able to place fire on the enemy.

0630 hrs. —At this time it became obvious that: (1) the enemy was trying to effect a withdrawal, and (2) he had waited too long and had a large number of men trapped inside the perimeter. As the light conditions improved, the action took on the characteristics of a turkey shoot. I believe without a doubt the enemy took his greatest casualties during this period. He simply waited too long.

0730 hrs. —Action terminated except for some small pockets of resistance. USMC elements immediately started sweeping the area outside the perimeter. MEDEVAC operations commenced and continued throughout the remainder of the morning.

Captain Chamberlain concluded his report with the comment: "The overall performance of the CIDG troops was quite commendable. They responded well to orders issued by the company commander. The camp commander . . . was a great help in stabilizing the men with his calm, don't-get-excited attitude."

The rapid reaction capability of the mobile strike forces in camp defense and in support of camp operations is illustrated by the following action. From January through March 1967, repeated contact with the enemy was made in the vicinity of Camp Bu Dop, Phuoc Long Province, in III Corps. On 14 January one CIDG company from Bu Dop, accompanied by two U.S. Special Forces and two Vietnamese Special Forces men, departed the camp with the mission of conducting a search and destroy operation with reconnaissance in Bu Dop's tactical area of responsibility. The patrol left at 0400. The terrain consisted of rubber trees, thick undergrowth, bamboo thickets, and savannah grass. The path of the patrol ran through an area where a reconnaissance patrol from Bu Dop had encountered the enemy on 8 December 1966. As the patrol drew closer to that area, the men heard shots and the company immediately broke into three files and started to search the area. At 1155, the point squad received fire and suffered one killed and one wounded. Firing broke out all around the company, and advisers Sergeant First Class R. Williams and Staff Sergeant J. Boorman made radio contact with a forward air controller who requested air strikes. The company was receiving heavy fire from the right flank, and seven more soldiers were wounded and two killed. Williams and Boorman advised an assault from the right flank to prevent encirclement. The company was still in its initial assault, but a momentary lull enabled the men to follow this tactic. The CIDG company attacked and overran the enemy positions, and the disorganized enemy withdrew in disarray. At 1220 an air strike was made on the retreating enemy. The troops continued the assault, finding the bodies of twenty-five North Vietnam Army soldiers in the enemy positions. By this time, the company had suffered three killed and eleven wounded. Under cover of air strikes, the company withdrew to the west to locate a landing zone for medical evacuation. While waiting for the helicopters the men collected and assembled all equipment and documents captured. The company was still receiving small arms fire from the east, and additional air strikes were made in that direction. The landing zone was secured. and medical evacuation was completed by 1500. At 1730 Captain Chester Garrett, with an additional company, landed and immediately started searching the area. The first company was extracted at 1730 and returned to base camp. While searching the area, Garrett found that an additional sixteen North Vietnam Army soldiers had been killed. Twenty-five maps and schematic drawings that were also found indicated that plans were being made to attack Camp Bu Dop and Bo Duc Subsector. At 1915 the relief force withdrew to base camp, arriving there at 2245. Within two hours after the first contact was made in this operation, the mobile strike force at Nha Trang had been airlifted to Bu Dop to defend the camp, while Captain Garrett took the second CIDG company in to relieve the company which had made the initial contact.

At 0745 on 17 February 1967, the 1st Company of the Third Nung Battalion (Airborne), III Corps mobile strike force, arrived at Bu Dop to reinforce the camp and conduct offensive operations to the east of camp. On 20 February, this company made contact with more than a battalion from the 12th North Vietnam Army Regiment, which was armed with 57-mm. recoilless rifles, 82-mm.

mortars, and the standard enemy small arms. The North Vietnamese directed an extremely heavy volume of fire into the advancing troops, but the civilian irregulars and the Special Forces troops, using sound infantry tactics, outmaneuvered the enemy and gained fire superiority. Fighting continued until tactical aircraft attacked. The enemy broke off the fighting and headed for the Cambodian border. Withdrawing south to a landing zone where medical evacuation was accomplished and reinforcements were brought in, the company then moved west back into an area protected by Bu Dop's artillery unit to rest for the night. The result of this action was 40 North Vietnam Army soldiers killed, 1 Special Forces soldier killed, 1 CIDG soldier killed, and 7 CIDG soldiers wounded.

On 23 March 1967, two CIDG companies from Bu Dop engaged a reinforced company of the North Vietnam Army approximately ten kilometers east of camp. Twenty of the enemy were killed in this action along with another estimated forty killed by U.S. air strikes that were requested. On 24 March, a CIDG company and a mobile strike force company conducted a heliborne assault on the same area and shortly thereafter became heavily engaged with two North Vietnam Army battalions, equipped with automatic weapons and recoilless rifles and supported by mortars. Casualties were 3 CIDG men killed and 11 wounded and 2 Special Forces men and 14 mobile strike force soldiers missing. It was confirmed that 98 of the enemy were killed, with an estimated 170 killed by air strikes.

In IV Corps a significant incident took place in the area of Camp Thuong Thoi, A-425, Kien Phong Province, on 6 January 1967. A company-size search and destroy patrol engaged an estimated Viet Cong company in fortified positions. A-425 requested assistance from the mobile strike force to hold the Viet Cong in position and destroy them before they could slip across the Cambodian border. A U.S. Air Force forward air controller adjusted artillery fire and called for an immediate air strike while the mobile strike force was being committed. Within three hours of the request, enough helicopters had been assembled at Can Tho, and the mobile strike force company had been lifted from there into position, but not before the Viet Cong company was able to withdraw into Cambodia under cover of darkness and carry away the dead and wounded. All indications were that the Viet Cong suffered heavy casualties.

Besides their participation in operations originating in the CIDG camps, the mobile strike forces also engaged in a large number of operations of their own in each of the four corps tactical zones, which were the mobile strike forces' operational areas. The strength of the mobile strike forces was doubled in the period from June 1966 through June 1967. The Nha Trang Mobile Strike Force, B–55, later the 5th Mobile Strike Force Command, was under the group commander's control. It grew to reinforced battalion strength in three years and saw action in all four corps tactical zones. Originally organized in 1964 under unilateral Special Forces command, the mobile strike forces were brought under joint U.S.–Vietnamese Special Forces command in December 1966. As of mid-1967 the Special Forces operations base at Nha Trang and the C detachment, Company A, in III Corps, each had five mobile strike force companies. Companies C, B, and D in I, II, and IV Corps, respectively, had three mobile strike force companies each. Mobile strike forces were flexibly organized, but a typical company had a headquarters, three rifle platoons, a weapons platoon, with mortar and machine gun sections, and a reconnaissance platoon. (*Chart 6*)

Mobile strike operations were conducted by irregular forces specially organized, trained, and equipped to rove the enemy rear for extended periods of time, conducting reconnaissance in force; to seek out and raid enemy bases; to interdict enemy lines of communication and support; to ambush and, if possible, to pin down and destroy small enemy units; and to establish contact with the enemy's large units so that major air and ground forces could be called in to destroy them. As developed by the Special Forces in Vietnam, the concept of mobile irregular warfare in support of counterinsurgency operations was predicated in part on the availability of tactical and strategic airpower and of an organized force of at least battalion strength with organic or provided transportation resources for immediate commitment to major engagements initiated by the mobile irregulars. Many combat actions were initiated by mobile strike forces during 1966 and 1967. On 2 April 1967 at 0030, patrols from the IV Corps Mobile Strike Force Company providing local security for Camp My An, A–426, detected an estimated 300 Viet Cong approximately three and half kilometers south of the camp. The Viet Cong were taken under fire, and the mobile strike force patrols fought their way back to defensive positions where the remainder of the mobile strike force company also took the Viet Cong under fire. Gunships and an AC–47 aircraft armed with automatic weapons were called in for support. Firefights lasted approximately three hours and included hand-to-hand combat. The Viet Cong broke away at 0330 and withdrew to the southwest. Results of the action were 1 Special Forces man wounded, 6 mobile

Chart 6—Organization of Mobile Strike Force Company

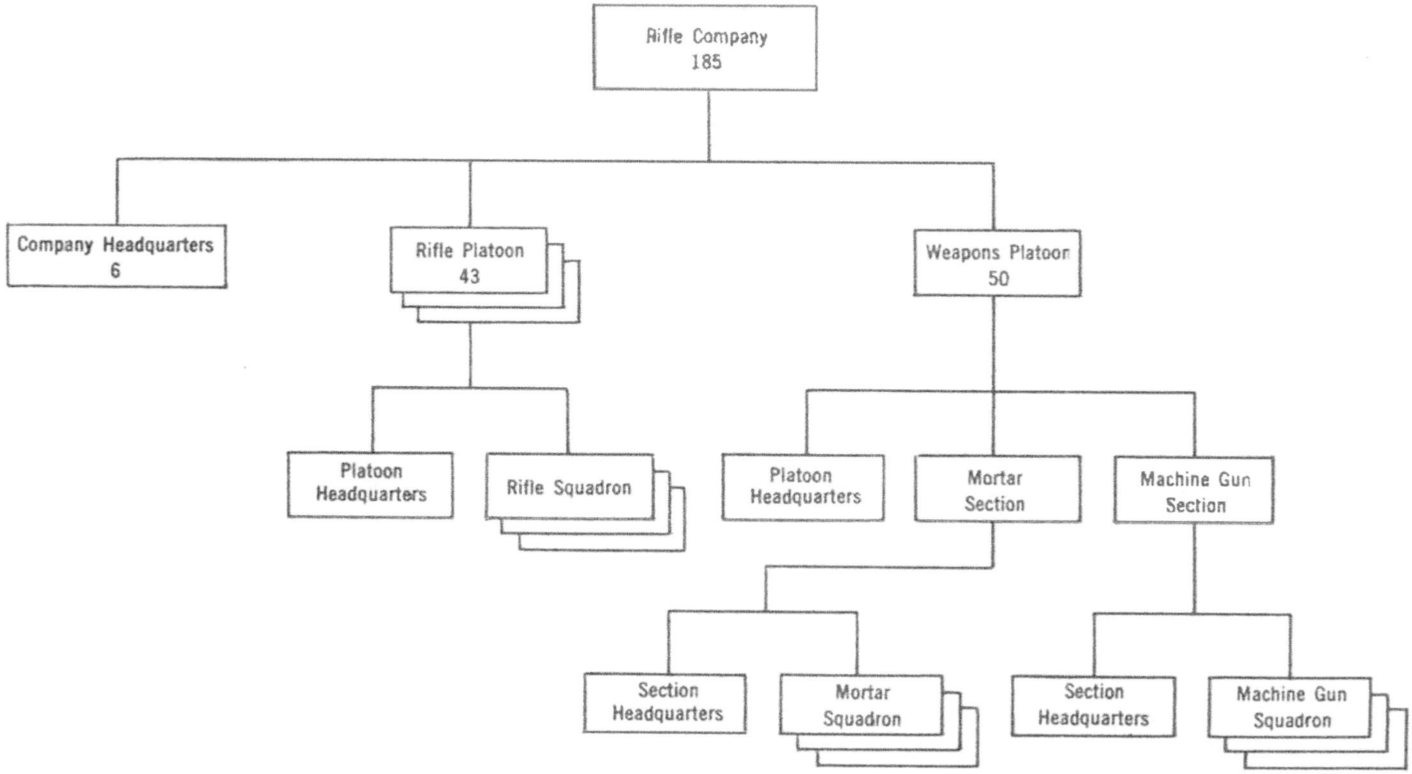

strike force men killed, 22 mobile strike force men wounded, and 2 carbines lost. The enemy lost 73 killed and eight small arms and assorted ammunition and supplies captured. At 1500, a new mobile strike force company was moved from Don Phuc, A-430, and relieved the company at My An. Company D, IV Corps, recommended the mobile strike force company for an award for its outstanding action against the Viet Cong battalion.

In III Corps Tactical Zone the most significant action of the quarterly period ending 31 January 1967 occurred near Camp Suoi Da, A–322, in Tay Ninh Province. On 2 November III Corps mobile strike force troops moved by air into the Suoi Da area to conduct reconnaissance missions against a known Viet Cong base area. Eight separate fights with elements of several Viet Cong regiments took place, and the mobile strike force was credited with rendering a Viet Cong battalion ineffective. Total confirmed enemy casualties were 85 killed, but another 148 were probably killed in ground action. This action was one of many in a situation that later led to the launching of Operation ATTLEBORO, up to that time the largest of the war involving conventional U.S. and Vietnamese units. (*See Appendix E.*)

Another significant area of mobile strike force employment was in the conduct of special operations. As part of a special operation named BLACKJACK 41C, three companies of the Nha Trang Mobile Strike Force conducted a parachute assault in the Seven Mountains region of IV Corps in May 1967. BLACKJACK 41C grew directly out of the special operation BLACKJACK 41, which had the mission of conducting unconventional warfare against the Viet Cong forces concentrated in the Seven Mountains region—a region which had long been a notorious Viet Cong stronghold. In addition to mobile strike forces, CIDG companies and a mobile guerrilla force participated in BLACKJACK 41C. The mission of BLACKJACK 41C was to exploit the intelligence gathered by the IV Corps Company D Mobile Strike Force and conduct operations to seal off Nui Giai Mountain in the Seven Mountains, denying the Viet Cong further use of it: to inflict maximum casualties on Viet Cong forces in the area; and to destroy Viet Cong command and supply installations.

The Company D Mobile Strike Force had been operating on Nui Giai Mountain since 20 April. On 9 May 1967 it received intelligence that one company from the 512th Viet Cong Battalion was operating on Nui Giai and would defend the mountain. It was decided to launch a large-scale operation to seal off Nui Giai and clear it of Viet Cong. The operation began at 0645 on 13 May with thirty-nine Special Forces troops and three companies of the 5th

Special Forces Group Mobile Strike Force conducting an airborne assault on drop zone BLACKJACK, south of Nui Giai. The drop took place from five C-130's flying in close formation (in 2,000-foot trail) and with the paratroopers jumping from a 700-foot altitude. No resistance was encountered in the drop zone. A mobile guerrilla company moved from Chi Lang at 0001 on 13 May and assumed a blocking position on the southeastern tip of Nui Giai. Elements of B-42, three CIDG companies from Vinh Gia, Ba Xoai, and Tinh Bien, assumed blocking positions on the north, west, and northeast of Nui Giai. During the first three days of the operation, the blocking forces conducted probes from the south of Nui Giai. On the fourth day of operations all elements began deep probes into the mountain. Contact was made with the Viet Cong every day except on 15 May. The 5th Group Mobile Strike Force and the mobile guerrilla company made repeated contact with an estimated two platoons of Viet Cong armed with two light machine guns, a 60-mm. mortar, Browning automatic rifles, and assorted small arms. The Viet Cong were in fortified, camouflaged positions. A total of seven air strikes were conducted on this location. The Company D Mobile Strike Force was evacuated from the mountain on 16 May and the operation terminated on 18 May. One company from Vinh Gia and one company from Tinh Bien remained on Nui Giai to conduct further operations and deny the Viet Cong use of the mountain.

As a result of the action, it was confirmed that 40 of the enemy were killed; 12 were wounded. Of the friendly forces, 9 were killed—8 of them CIDG men and 1 man from the strike force; 35 were wounded—12 CIDG men, 4 men of the strike force, and 9 men of the U.S. Special Forces. Eleven small arms were captured as well as medicines, assorted documents, and some field equipment. Three thousand pounds of rice, ten squad-size huts, fifteen caves, and one command post were destroyed.

The biggest single lesson learned from the campaign was that a large-scale operation like BLACKJACK 41C, involving eleven companies and an airborne assault marshaled out of Nha Trang, could be assembled and launched in a very short time. The operation was conceived, planned, and put into effect in less than thirty-six hours. Only eighteen hours elapsed from the time the operations order was issued to all task force commanders at Can Tho until the first elements were dropped on the BLACKJACK landing zone.

In concluding this section on the events of June 1966–June 1967, some comment is appropriate on 6th Special Forces Group

CANDY AFTER MEDICINE

developments in the fields of training, revolutionary development activities, and logistics.

In conjunction with its CIDG and special operations missions. the group expanded its training activities at the Dong Ba Thin Training Center, a key element in the constant effort to upgrade the competence and efficiency of Vietnamese Special Forces and CIDG troops to the point where they could be self-sustaining. During 1966 and 1967 there was a significant increase in the quality, capacity, and diversity of leadership training and advanced Special Forces courses presented at the center. A large part of the instruction was being given by Vietnamese Special Forces by the summer of 1967. In order to have all mobile strike forces qualified as parachutists, a basic airborne school was established at the center. The planned input was one CIDG company of 132 persons per month, but the school had the capacity to qualify one battalion per month if the need arose.

On 15 September 1966, by direction of General Westmoreland. the 5th Special Forces Group established the Military Assistance Command, Vietnam, Recondo School. The school trained selected troops from all Free World Military Assistance Forces in the long-range reconnaissance techniques developed and employed by Project Delta. The school was able to conduct the following training: a three-week reconnaissance-commando course, with a maximum school capacity of 120 students—a class of 60 students began training every two weeks; a one-week airborne course for reconnaissance-commando students; mandatory subject training such as escape and evasion and survival; a one-week long-range reconnaissance patrol course for men assigned to Projects Delta, Sigma, and Omega; and special training such as high altitude–low opening parachuting, scuba, or other training missions as required. Selected Vietnam Army cadre were trained also as the nucleus of the Vietnam Army Reconnaissance-Commando School, which was scheduled for establishment.

Developed in 1966, the concept of revolutionary development included civic action, troop motivation, and psychological operations. In the course of 1966–1967 the 5th Group unified its efforts in these areas under a new revolutionary development support activities section at the 5th Group headquarters in Nha Trang, which had the responsibility of planning, supervising, and evaluating U.S. Special Forces participation in the revolutionary development program. The major point of emphasis was that revolutionary development had to be a Vietnamese program, and in the final analysis its success or failure depended on the Vietnamese. If the people were to be won over to the government, the revolutionary development project had to be Vietnamese—a fact which was unquestionably borne out by events after 1961. Medical aid continued to be the most popular and significant form of civic action. Finally, the Civic Action Guide published by the 5th Group pointed up the need to develop projects that could be supported locally and continued by the people after U.S. and Vietnamese Special Forces troops left the area.

The unusual logistic demands of the Special Forces and CIDG program continued, with the added requirements of greater expansion. Forward supply points in each of the four corps tactical zones, heavy emphasis on aerial resupply, and rapid resupply techniques continued to be used and expanded. Two new logistic requirements that arose in the course of the year dealt with air support for guerrilla operations and the logistical needs of the prestocked fighting camps. To speed resupply, the so-called red ball system was established. Designed to provide rapid response to urgent supply requests, the red ball made resupply within one hour a reality. High-priority requests received a distinctive red ball marking that insured special handling from the time a request was received until the item was pulled from the warehouse, placed on line, and loaded on aircraft. From the time of its inception on 15 March 1967 to the end of May, over 115 U.S. Special Forces red ball requisitions were processed.

The entire logistic support system for the U.S. Special Forces and CIDG program in Vietnam was exceptionally efficient. The whole countrywide operation was handled by 250 U.S. soldiers, eighty of whom were Special Forces men. All told, the 5th Group developed what was considered one of the most efficient and economical supply systems in Vietnam. Operational control was exercised differently in each of the corps areas. (*Chart 7*) This was especially true in regard to operations conducted in conjunction with conventional forces.

The Tet Offensive: The CIDG Grows Up

Three major changes in the Special Forces Civilian Irregular Defense Group effort took place from June 1967 to June 1968 and in many respects set the stage for the remainder of Special Forces participation in the war in Vietnam. Two of these were the result of policy changes which came from within the group, while the third change came about as a result of the enemy's 1968 *Tet* offensive.

The first change was a significant increase in so-called Vietnamization. In early 1968, senior commanders of the Military Assistance Command, Vietnam, were discussing methods of increasing the number of Vietnamese involved in the conflict, reducing the number of American casualties, and releasing U.S. units for responsive deployment by U.S. Military Assistance Command, Vietnam. In essence, the concept which developed and which was adhered to through 1970 called for the CIDG camp and mobile strike forces to be employed as balanced forces to establish an interdiction zone in the western border area of South Vietnam. The 5th Special Forces Group began to prepare for disengagement with the development of plans for the assimilation of the CIDG by the Vietnam Army or other government agencies. This included renewed emphasis on the closure, turnover, and conversion of CIDG camps. Within the group itself, a number of steps were taken to lay the groundwork for Vietnamese take-over of the CIDG program. A new program was initiated to improve Vietnamese counterpart relations and stress the training of the Vietnamese so that they could accept

Chart 7—Command Structure of 5th Special Forces Group (Airborne)

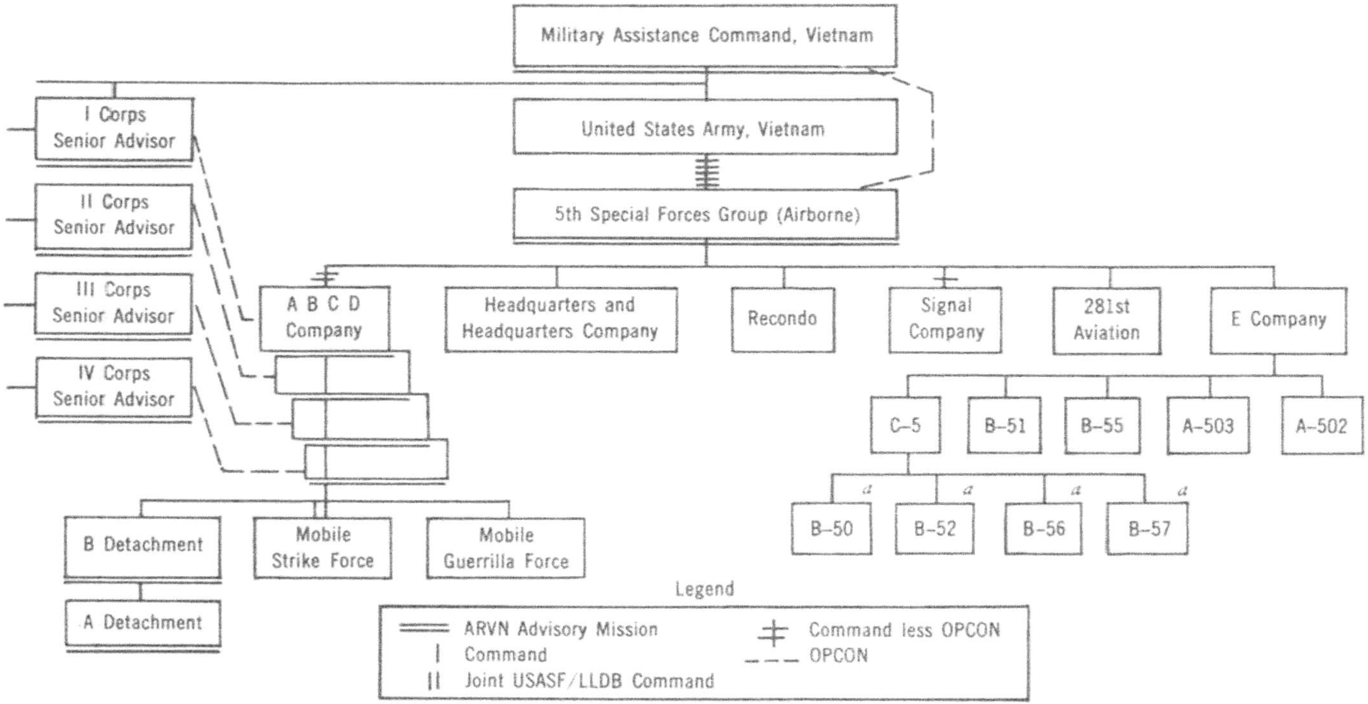

[a] May be placed under OPCON of Corps Senior Advisor, Field Force Commander, or Commanding General of III Marine Amphibious Force, on order of COMUSMACV.

greater responsibilities in the direction of the CIDG program. Extensive administrative, operational, and logistical responsibility was delegated to U.S. Special Forces company commanders, one in each of the four corps tactical zones, and they were encouraged to establish and maintain the closest relationship possible with the senior U.S. and Vietnamese commanders in their respective zones. Within the group headquarters itself, staff sections were oriented toward a mission of co-ordination, support, service, and planning, which included making long-range studies for future commitment. Finally, to the extent possible, assignments were lengthened so that U.S. Special Forces men could spend a longer time with their Vietnamese counterparts to increase the operational efficiency of the Vietnamese.

The second change in the Special Forces CIDG effort was in the area of revolutionary development, civic action, and psychological operations, and this too grew out of the increasing emphasis on Vietnamization. The mission of the group in the civic action and psychological operations field was to advise the Vietnamese Special Forces, to support the government revolutionary development program, and to conduct psychological operations against the Viet Cong. Because of the difference in organization and responsibilities of the group S–5 section and its counterpart, the political warfare section of the Vietnamese Special Forces high command, however, this mission was not being fully achieved. In the political warfare section, the officer in charge was responsible not only for the civic action and psychological operations but also for the motivation and indoctrination of the indigenous troops—their morale and welfare. Therefore, in order to provide compatible advisory assistance to the Vietnamese political warfare section, the 5th Special Forces Group psychological operations section was moved from the S–3 to the S–5 section, and the combined civic action–psychological operations section was augmented with a political warfare section that included a motivation and indoctrination officer. This officer was responsible for education, information, troop benefits, dependent care, and other matters affecting the morale and welfare of the civilian irregular troops. Along the lines of the Vietnamese political warfare concept, the first priority of the effort went to the troops, the next to the population, and the next after that to dealing with the enemy. Great emphasis was given to providing assistance to the CIDG soldier and his family. The net result of this reorganization of the U.S. Special Forces civic action and psychological operations effort was to increase Vietnamese participation, thereby preparing the Vietnamese for total take-over and at the same time

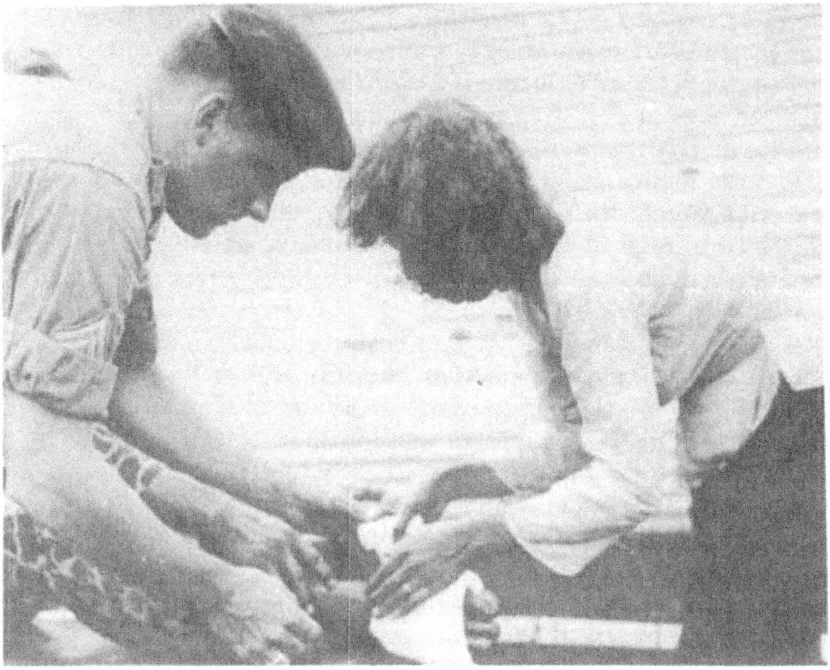

TEACHING WIFE OF VIETNAMESE SOLDIER TO TREAT HIS FOOT WOUND

improving the motivation of the CIDG for conversion to complete Vietnamese control.

Finally, the third change in the Special Forces CIDG effort that occurred during this year came as a response to the enemy *Tet* offensive and the change in the nature of the war which it brought about. In the course of the *Tet* offensive, CIDG troops were employed in the defense of certain urban centers, a combat role new to them but one in which they proved to be very effective. Further, after *Tet*, the CIDG came to be regarded as an economy of force element which could be used to release conventional units for deployment in response to new enemy buildups. These tactical and strategic changes in the employment of civilian irregulars reflected once again the flexibility and responsiveness of the U.S. Special Forces CIDG effort and were further evidence of the Special Forces' wide-ranging counterinsurgency capabilities.

With the large-scale introduction of conventional North Vietnamese Army forces and sophisticated enemy weapons into South Vietnam, the CIDG program was re-evaluated to analyze its effectiveness in the light of the transition from an insurgency situation to one of insurgency coupled with "hot" war. During the twelve months major changes in enemy armament occurred. Introduced in quantity were tube artillery, large rockets, large mortars, modern small arms of the AK47 type, antiaircraft artillery up to 37-mm., and heavy machine guns. Tanks were employed on one occasion against the CIDG camp at Lang Vei, and others were sighted in Laos and Cambodia near the border and in South Vietnam. In central and southern South Vietnam, North Vietnamese Army replacements were used to bolster main force Viet Cong units that had lost many men.

The enemy launched his *Tet* offensive on 29 January 1968. This was followed by a massive buildup at Khe Sanh and the armor-supported attack that overran the camp at Lang Vei in I Corps. Pressure on CIDG camps, except for the attack on Lang Vei, was unusually light during the entire *Tet* offensive and for approximately sixty days thereafter. As the enemy withdrew from the vicinity of the urban areas, pressure on the CIDG camps increased, principally in the form of frequent attacks by mortar and rocket fire on camps near known enemy infiltration routes and base areas. CIDG forces responded in an exemplary manner in all corps tactical zones during *Tet* and were responsible for the successful defense of several urban areas, as well as the rapid relief of others. Among the urban centers successfully defended by the CIDG were Nha Trang, Qui Nhon, Kontum, Pleiku, Chau Doc, Ban Me Thuot, Phan Thiet, and Dalat. The street-fighting ability demonstrated by the CIDG troops in the defense of these towns was somewhat surprising in view of the fact that their training had not been geared for that kind of combat, but their superior performance demonstrated conclusively that the CIDG soldier was the combat equal of any soldier in Vietnam. Immediately following *Tet*, a major tactical redeployment of conventional troop units was necessary in order to provide forces to counter the increased enemy threat in northern I Corps. It was at this point that CIDG troops were used as economy of force elements in order to make this redeployment of conventional units possible.

The effects of the *Tet* offensive were also felt in the effort to turn over the CIDG program to the Vietnamese. In many cases the schedule for the turnover or conversion of certain camps was brought to a virtual standstill by the *Tet* offensive. Areas thought to be secure and ready for conversion or turnover proved not to be in the light of *Tet*, and the schedules had to be revised.

Besides these three major changes in the character and conduct of the U.S. Special Forces CIDG program, there were other developments which deserve mention. The superior performance of the mobile strike forces prompted the authorization for a total of 47 mobile strike force companies—up 28 companies from the number authorized the Special Forces during the 1966–1967 period. By the summer of 1968, 34 mobile strike companies were in existence: 5 in I Corps, 12 in II Corps, 7 in III Corps, and 10 in IV Corps. Construction of new camps, while not nearly as widespread as in 1965 and 1966, continued. While the *Tet* offensive did slow it down, the turnover process continued with the successful turnover to complete Vietnamese Special Forces control of three camps. In response to the increase in enemy firepower and in recognition of the valuable tactical role played by the CIDG, a weapons modernization program was submitted to and approved by Military Assistance Command, Vietnam, in April 1968 under which the CIDG troops were given priority in the receipt of M16 rifles, M60 machine guns, and M79 grenade launchers. Up to this point, the CIDG troops had been using M1 carbines and some M14's. As the new weapons became available, they were issued to the mobile strike forces, the combat reconnaissance platoons, and the camp strike forces, in that order. The weapons transfer program was completed by January 1969, and the combat effectiveness of the CIDG was significantly increased as a result. The use of both camp and mobile strike forces in conventional operations in conjunction with U.S., Vietnam Army, and Free World forces took place on an ever-increasing scale, presaging the eventual assimilation of the CIDG into the Vietnamese military organization.

An enemy assault on Camp Loc Ninh, A–331, in III Corps, took place from 29 October to 4 November 1967. Although the assault was a determined enemy attempt to overrun a camp, the camp strike force, together with elements of the 1st U.S. Infantry Division which reinforced it on the second day, successfully defended the camp with no outside help except air strikes, and dealt the enemy an extremely heavy defeat in which he lost over 1,000 killed. Of that figure, 184 enemy killed were credited to the civilian irregulars and the U.S. and Vietnamese Special Forces at Camp Loc Ninh. Against that, the camp casualties in the action were 4 Special Forces men wounded, 6 CIDG men killed, and 39 CIDG men wounded. The following account of the action at Loc Ninh is paraphrased from the operational report for the quarterly period ending 31 January 1968.

Before the attack, one CIDG company was engaged in patrolling approximately eight kilometers northwest of Loc Ninh, while three

Civilian Irregulars Learning To Fire a Machine Gun

companies, a reconnaissance element, and a civic action and psychological operations squad were inside the camp. The CIDG camps were organized for defense, with the troops billeted in the wall of the defensive perimeter and each CIDG company assigned a specific part of the perimeter and the responsibility for defending it. All basic weapons were in defensive positions at all times.

Since mid-September 1967, reports from significant problem areas and intelligence reports of the Free World military forces, the Vietnam Rangers, and the Special Forces had all indicated that the Viet Cong were making preparations for a large-scale ground assault in Loc Ninh District. In early October allied intelligence placed the time of the attack between 22 and 30 November. Enemy reinforcements were observed moving into Loc Ninh District. The 272d Viet Cong Regiment, which had been located in the Bu Dinh Secret Zone, moved into the vicinity of Loc Ninh in early July and remained there except for making one excursion to the south. The 273d Viet Cong Regiment was observed moving north in mid-August from its previous position in War Zone C to a position near Dong Xoai. One week before the attack on Loc Ninh, the regiment crossed the Song Be and took its position northwest of Loc Ninh; it became the camp's primary Viet Cong protagonist. The 165th North Vietnamese Army Regiment operated in the fishhook area after its 7 August attack on Camp Tong Le Chon and was believed to have moved farther northeast into Loc Ninh District in early November. At least one battalion and possibly two battalions of the 165th Regiment participated in the attacks on Loc Ninh. The 84A North Vietnamese Army Artillery Regiment was believed to have moved some attack forces into northwest Phuoc Long in mid-October, while the 141st North Vietnamese Army Regiment is not believed to have moved any substantial units into the battle area, though troops from the 141st Regiment may have been assigned to other attacking units. Captured documents indicated that a few of the enemy killed in action belonged to the 141st Regiment. After the 273d Viet Cong Regiment moved north, the 9th Viet Cong Division headquarters moved from War Zone D to Loc Ninh in the latter part of September 1967 and appears to have supervised the attack on Loc Ninh, which began on 29 October 1967 at 0115. All the aforementioned intelligence was known before the attack. The actual extent in numbers of the attackers and the duration of the attack were unknown. There were no indications that a prolonged and fanatical attempt would be made against the CIDG camps in Loc Ninh District.

During the first hours of 29 October the subsector compound north of Loc Ninh began receiving 82-mm. mortar and heavy small arms and automatic weapons fire from the northwest. Within a few minutes of each other, Camps Loc Ninh and Hon Quan began receiving mortar fire in and around the compound. Loc Ninh received continuous heavy mortar fire until approximately 0250 after which it became sporadic until it ceased at 0530. During the initial heavy mortar attack, Loc Ninh Special Forces Camp A–331 received approximately twelve rounds inside the compound, with five rounds landing in the vicinity of the gate to the camp. At Hon Quan no rounds landed within the compound. Hon Quan received approximately sixty rounds of 82-mm. mortars during the first hour, and the province chief's house was the target of some eight to twelve 57-mm. recoilless rifle rounds. At 0115 the Viet Cong struck the subsector of the Regional Forces and Popular Forces compound with co-ordinated mortar and ground attacks, and at 0220 after an hour of fighting it was reported that Viet Cong were within the compound. U.S. troops could not substantiate the report until 0520 when it was confirmed that one U.S. Regional

Forces and Popular Forces adviser and a district chief were in the command bunker within the compound.

For four and a half hours under cover of darkness, an estimated battalion of the 273d Viet Cong Regiment supported by a battalion from the 84A North Vietnamese Army Artillery Regiment attacked Camp Loc Ninh with mortar, rocket, heavy machine gun, and small arms fire. The camp received an estimated one hundred and eighty 82-mm. and 120-mm. mortar rounds and fifteen RPG40 rounds. The camp went on full alert at 0115. Forward air controllers, Spooky (C-47 aircraft), light fire teams, and tactical air support were at their stations within fifteen minutes and gave continuous support to Loc Ninh throughout the night until the enemy broke contact at 0520. At this time detachment A–331 at Loc Ninh launched an operation with two CIDG companies to relieve the Regional Forces and Popular Forces compound. By 1600 the compound had been secured and all Viet Cong expelled.

During the attack Loc Ninh carried out a field operation in which twenty-three men of the Viet Cong and North Vietnamese Army were killed.

After the attack ceased at 0530 on 29 October the Loc Ninh Special Forces Camp immediately began to improve its defensive position. On 31 October at 0050 the camp was again attacked by an estimated two battalions of the 273d Viet Cong Regiment, supported by a battalion from the 84A North Vietnamese Army Artillery Regiment. It was estimated that the camp received two hundred rounds from 82-mm. and 120-mm. mortars and eighteen rounds from 122-mm. rockets, as well as RPG40, RPG7V, and recoilless rifle fire of undetermined caliber. A Viet Cong battalion attempted a mass assault on the camp, but the attack was broken up and the Viet Cong were pinned down and then destroyed by coordinated fire from the camp and tactical air strikes. At the first sign of light the enemy withdrew to the north, northeast, and northwest. At 0200 on 1 November 1967 Camp Loc Ninh received approximately ten 82-mm. mortar rounds, believed to have been fired in order to allow the enemy to gather his dead and wounded from the battlefield. Contact with the enemy was light and sporadic until the following day at 0050 when the enemy again massed a battalion for an obvious last-ditch effort to overrun the camp. Once again the attack was repelled by the camp's withering defensive fire and especially well-placed air strikes. After the final air sortie, the enemy became disorganized and fled. Sporadic contact was maintained until dawn.

The enemy's main force was estimated at nine North Vietnamese Army and Viet Cong battalions, which participated in the attacks on Camp Loc Ninh from 29 October to 4 November 1967. The enemy units consisted of the following: 3 battalions from the 273d Viet Cong Regiment, which was believed to be located to the west of Loc Ninh; I battalion of the 84A Artillery Regiment deployed around Loc Ninh; 1 to 2 battalions of the 165th North Vietnamese Army Regiment located southeast of Loc Ninh; 2 to 3 battalions of the 272d Viet Cong Regiment located east of Loc Ninh; part of a battalion of the 141st North Vietnamese Army Regiment which was distributed among the other participating units; and a heavy weapons battalion equipped with antiaircraft guns and mortars. Each battalion, except for the heavy weapons battalion, is believed to have numbered between 300 and 400 men and to have been slightly understrength. The heavy weapons battalion organically has fewer men than an infantry battalion. Each infantry battalion was equipped with the usual Chinese or Soviet small arms and crew-served weapons. In addition, the 84A North Vietnamese Army Artillery Regiment provided 120-mm. mortars and 122-mm. rockets. One surface-to-air missile exploding in the air was observed by two U.S. forward air controllers. The missile was fired from an area approximately eight kilometers west of Loc Ninh.

The enemy employed no tactical innovations in his attacks on Loc Ninh. An attack usually commenced with heavy mortar bombardments, followed in quick succession by ground assaults that were preceded by squad-size sapper units coming from several directions. A larger attack consisted of several assault waves; during the height of battle Loc Ninh withstood at least three such full-scale ground assaults. Usually, the enemy's last offensive operation was a ruse for body recovery. Almost all Viet Cong operations were conducted under cover of darkness. The attackers usually began assembling after dusk and reached their offensive positions at about 2100. Most attacks were launched about 0030 and concluded at dawn, with scattered sniping in the early morning hours allowing the attack force an orderly withdrawal. Allied air reconnaissance observed heavier than usual antiaircraft guns surrounding Camp Loc Ninh. Some of the fire received at camp came from .50-caliber machine guns, some mounted as quads. While the enemy's antiaircraft fire was not effective, it did divert extensive suppressing fire from other targets. Recoilless rifles and antiaircraft weapons were employed from civilian homes, especially from multistoried structures. Despite the extensive preparations made by the enemy, the only compound they were able to penetrate was the Regional Forces compound on the first day of battle.

The discipline of the civilian irregular and Special Forces troops was outstanding during the assaults. As a result of periodic practice alerts, everyone knew exactly where to go and what to do. The troops of the U.S. Special Forces, Civilian Irregular Defense Group, and Vietnamese Special Forces with their commanders stayed with their assigned units throughout the attack, assisting wherever and whenever they were needed. The U.S. Special Forces detachment commander was in the communications bunker at first; after forwarding the required reports to higher headquarters, he took his position on the perimeter and directed the defense of the camp. The executive officer served at the point of greatest impact, assisted in resupplying the perimeter with ammunition, gave first aid, and helped with evacuation of the wounded and dead. The team sergeant was everywhere: he moved from position to position on the perimeter, offering encouragement and reassurance to the troops and forwarding necessary reports to the detachment commander. The team medic not only treated the wounded on the defensive perimeter but also in the medical bunker. Weapons men divided their time between the mortar crews and the perimeter. Team members were all periodically active and were exceptionally effective in keeping the camp defenses organized. Individual acts of heroism were too numerous to mention; suffice it to say that every U.S. team member was recommended for an award of valor.

There were additional contingency plans: camp defense plans and camp alert plans. Effective communications were maintained throughout the attack on Loc Ninh. Internal communication was excellent, as was external communication to higher headquarters. Even after the discovery that the outside antennas had been destroyed in the attack on Loc Ninh, communication was immediately regained by switching to the underground emergency antennas.

All requests to higher headquarters were met promptly. The flareships and Spooky were on the scene of battle within twenty-five minutes after their summons and they remained on the scene as long as they were needed. Forward air controllers, air and artillery support were outstanding. No requests for reinforcements were made to higher headquarters, but as the fight developed over the next several days, Vietnam Army units and the 1st Brigade of the 1st U.S. Infantry Division were airlifted into Loc Ninh. Logistical support was superb. Requests to higher headquarters for supplies and equipment were handled with immediate dispatch. Medical evacuation was swift once daylight came and the fighting abated. All reports were submitted to higher headquarters according to the standing operating procedure, and all deserving individuals were recommended for awards of valor.

Enemy pursuit, aside from the civilian irregular relief column dispatched to the Popular Forces compound, was left to the 1st U.S. Infantry Division in and around Loc Ninh. There were reports that the enemy was fleeing in all directions in a disorganized manner. Initially the enemy withdrew to the west of Loc Ninh, but heavy contact with U.S. and CIDG units diverted the Viet Cong to the north, northeast, and east. A trail survey later revealed that the enemy withdrew in battalion-size or larger units primarily northeast toward the Cambodian border and due east to the Bu Dinh Secret Zone.

A confidence bred of demonstrated ability created in the ranks of Civilian Irregular Defense Group troops a conviction that they could win against the Viet Cong and North Vietnamese regulars. With this conviction the transition to the offensive was complete.

CHAPTER VI

Unconventional Operations

With the formation of the mobile guerrilla force and Projects Sigma and Omega in the period August through October 1966, the ability of the Special Forces in South Vietnam to conduct unconventional operations was significantly increased. One of the chief characteristics of Special Forces units is their capacity to conduct brief or extended operations inside territory dominated or controlled by enemy forces. The conduct of these operations was one of the most significant contributions of the Special Forces to the war effort in Vietnam. Unconventional operations were planned and executed in furtherance of programs to fill three major needs of the government of South Vietnam and its supporting Free World Military Assistance Forces. The requirements were intelligence; denial, through harassment and interdiction, to the enemy of unrestricted use of various human and material resources essential to carrying out his strategic and tactical plans; and recovery of American, Free World forces, and Vietnamese soldiers missing in action.

Apart from *ad hoc* recovery attempts, the unconventional operations most frequently conducted in enemy-controlled areas were long-range reconnaissance patrols and mobile guerrilla actions. Fundamental to both types was the concept that, with proper training, organization, guidance, and support, soldiers who were indigenous to the area of operations would achieve the greatest success in locating enemy troops, bases, and auxiliary facilities. The concept of operations also held that, by virtue of the irregular status of the mission forces and their dependence on Special Forces advice, assistance, and special logistic and administrative support as organized in the CIDG program, unconventional operations were best conducted within the framework of that program.

Mobile guerrilla forces were created in the fall of 1966 in refinement and amplification of the mobile strike concept. These guerrilla units were organized, trained, and equipped to operate in remote areas previously considered to be Viet Cong or North Vietnam Army havens. Usually almost no reconnaissance or clearing operations had been carried out in such territory. Instituted as economy of force units, the troops of the mobile guerrilla forces would infiltrate an area to interdict enemy routes, conduct surveillance, seek out enemy forces and installations, and collect intelligence along their axis of advance. Viet Cong base camps were found, watched, and raided if possible, or were harassed if the enemy was too well defended and organized. Lines of communication were cut by means of raids and ambushes, and were planted with mines and booby traps. Storage areas for supplies were found and eliminated, and air strikes were directed and the results assessed.

A mobile guerrilla force unit was inserted into its assigned tactical area of operations by the most unobtrusive means available. Once in the area of operations, the unit became a true guerrilla force in every respect except that of living solely off the land. Selected items of resupply were delivered by air. The guerrilla force operated from mobile bases, and the troops were capable of remaining and operating in a particular area for thirty to sixty days. The guerrilla force required complete freedom of action within a specified area of operations in order to achieve success. For this reason, once an area was designated for the conduct of an operation, the mobile guerrilla force "owned" that area—including control of air support.

The guerrilla forces had essentially the same desirable characteristics as the mobile strike forces with the following exceptions: the mobile guerrilla force troops were highly responsive to the needs of the Special Forces companies in each of the four corps tactical zones in that operational control rested with the Special Forces company commander for the mobile guerrilla force located in his corps tactical zone; each mobile guerrilla force unit was wholly commanded and controlled by a Special Forces A detachment (the mobile strike forces went under joint U.S.–Vietnamese Special Forces command in December 1966); and each mobile guerrilla force unit was trained to operate as an independent unit with no reinforcement or mutual support.

The basic organization of the mobile guerrilla forces was the same as that of the mobile strike forces, with a 34-man combat reconnaissance platoon added as an organic unit. (*Chart 8*) The mobile guerrilla force unit was organized without a weapons platoon, but an M60 machine gun squad was included in the company headquarters. The combat reconnaissance platoon could be employed in advance of the mobile guerrilla force to provide reconnaissance, establish an initial resupply point, and gather intelligence. The combat reconnaissance platoon secured the patrol

Chart 8—Mobile Strike Force Organization

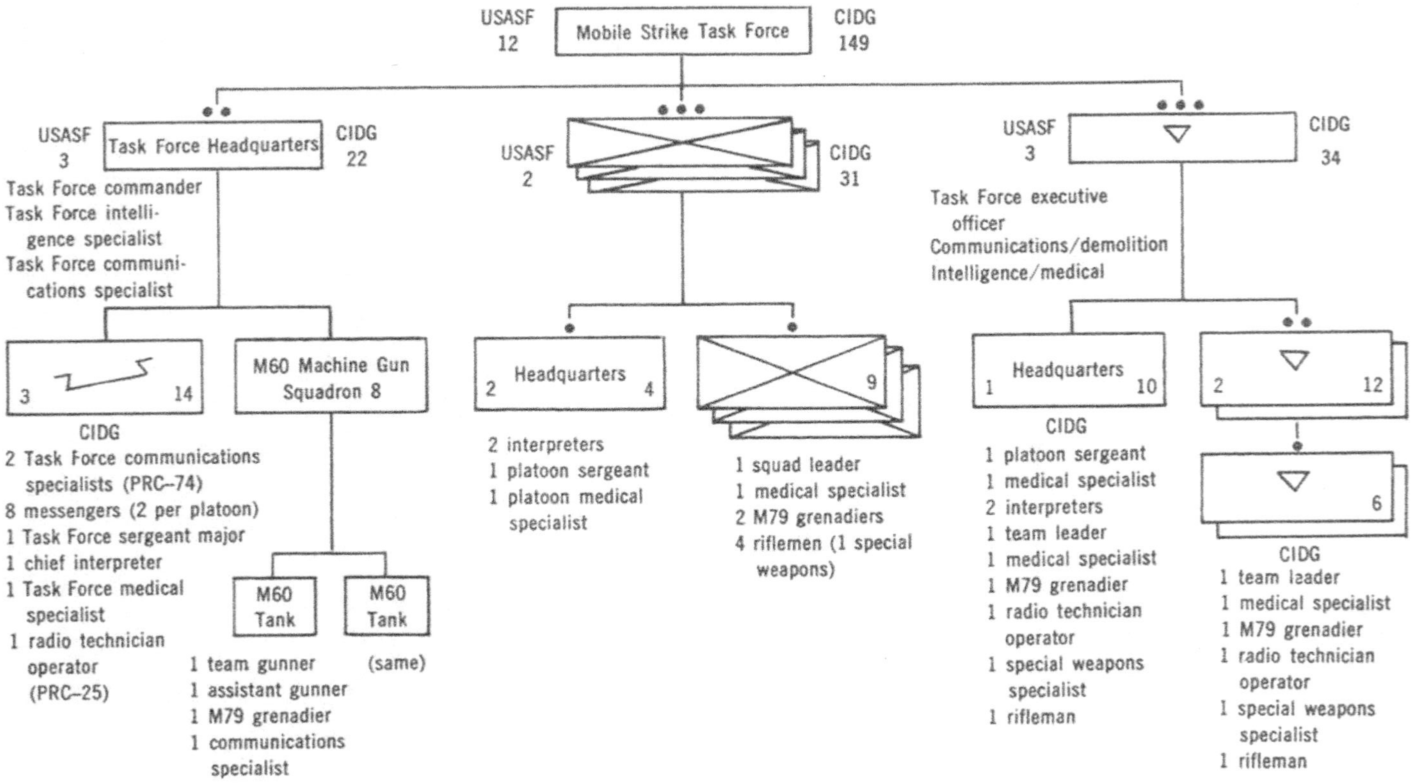

base and received the first resupply pending the arrival of the rest of the force. To avoid disclosing their position, mobile guerrilla force troops on many occasions were resupplied entirely through the use of modified, 500-pound napalm containers of pre-packaged, code-identified bundles delivered by A1E-type aircraft in what seemed to be a normal air strike.

Besides the development of one mobile guerrilla force in each corps tactical zone under the operational control of the Special Forces company commander for each corps, the 5th Special Forces Group (Airborne) also expanded its unconventional operations capability with the creation of Projects Omega and Sigma. The launching of Projects Omega and Sigma increased the long-range reconnaissance and intelligence-gathering capability of the 5th Group beyond that already furnished by Project Delta. Each of the projects had a strength of about 600 men plus an advisory command element organized as a modified B detachment. Each project consisted of a reconnaissance element and a reaction force. *(Charts 9 and 10)*

Though the strength and organization of these Greek-letter projects were similar, there were some important differences between them. Project Delta operated under joint U.S.-Vietnamese Special Forces command, was directly responsive to the requirements of Military Assistance Command, Vietnam, and the Vietnamese Joint General Staff, and had the Vietnam Army 91st Ranger Battalion (Airborne) assigned as its immediate reaction force. Projects Omega and Sigma, however, were commanded by the Special Forces, were respectively responsive to the requirements of I Field Force, Vietnam, and II Field Force, Vietnam, and had CIDG mobile strike force companies assigned as immediate reaction forces.

The following excerpts are taken from the group commander's debriefing report for the period June 1966 to June 1967 and describe the performance of Projects Delta, Omega, and Sigma during his year in command of the 5th Special Forces Group (Airborne).

Projects Omega (B-50) and Sigma (B-56) are under the OPCON of I FFORCEV and II FFORCEV respectively. Commanded directly by USASF personnel, these units are directly responsible to the requirements of the field force commanders. They are composed of 8 "Roadrunner" teams with four indigenous personnel each, and 8 reconnaissance teams with two USASF and four indigenous members each. The "Roadrunner" teams conduct long distance reconnaissance over enemy trail networks. The reconnaissance teams, on the other hand, conduct saturation patrols throughout specified reconnaissance zones, gathering

Chart 9—Organization of Project Delta

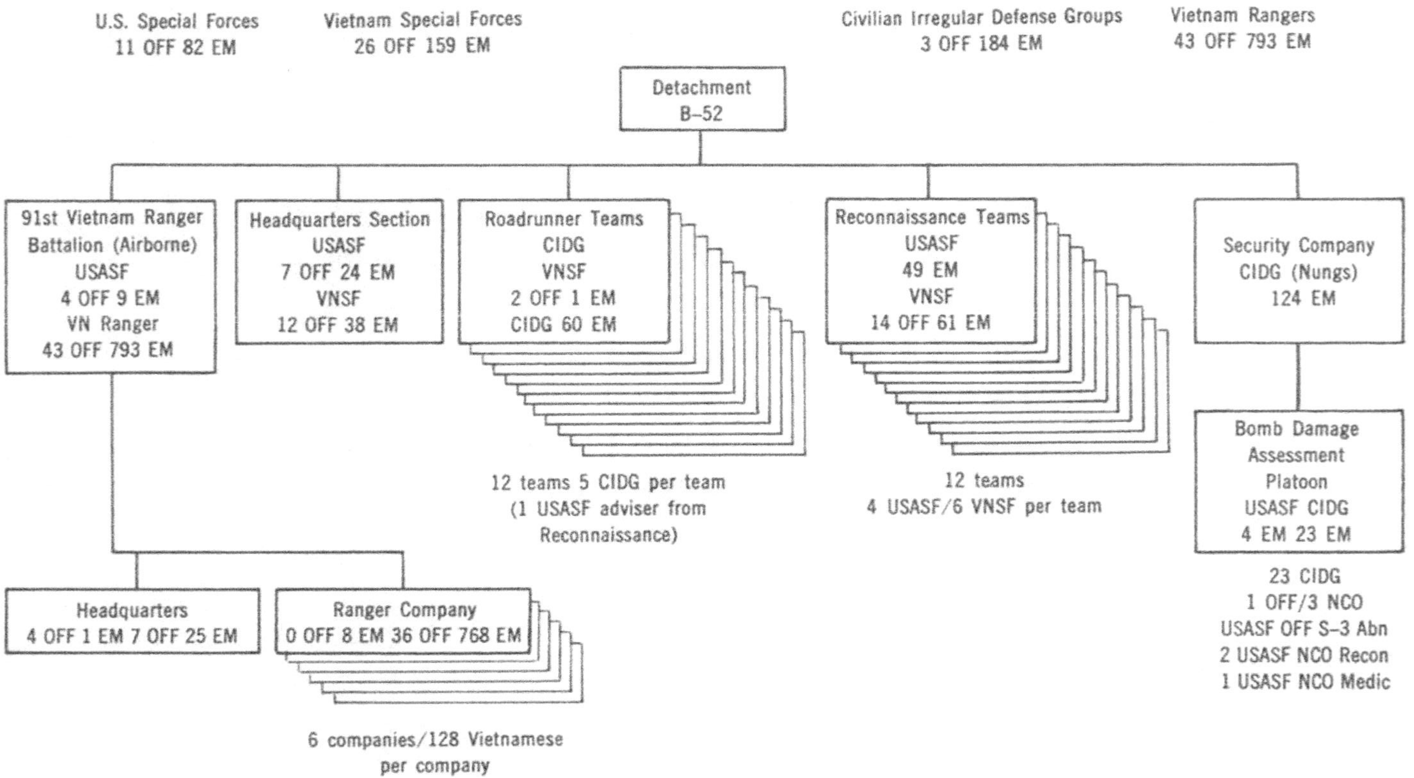

Chart 10—Organization of Projects Sigma and Omega

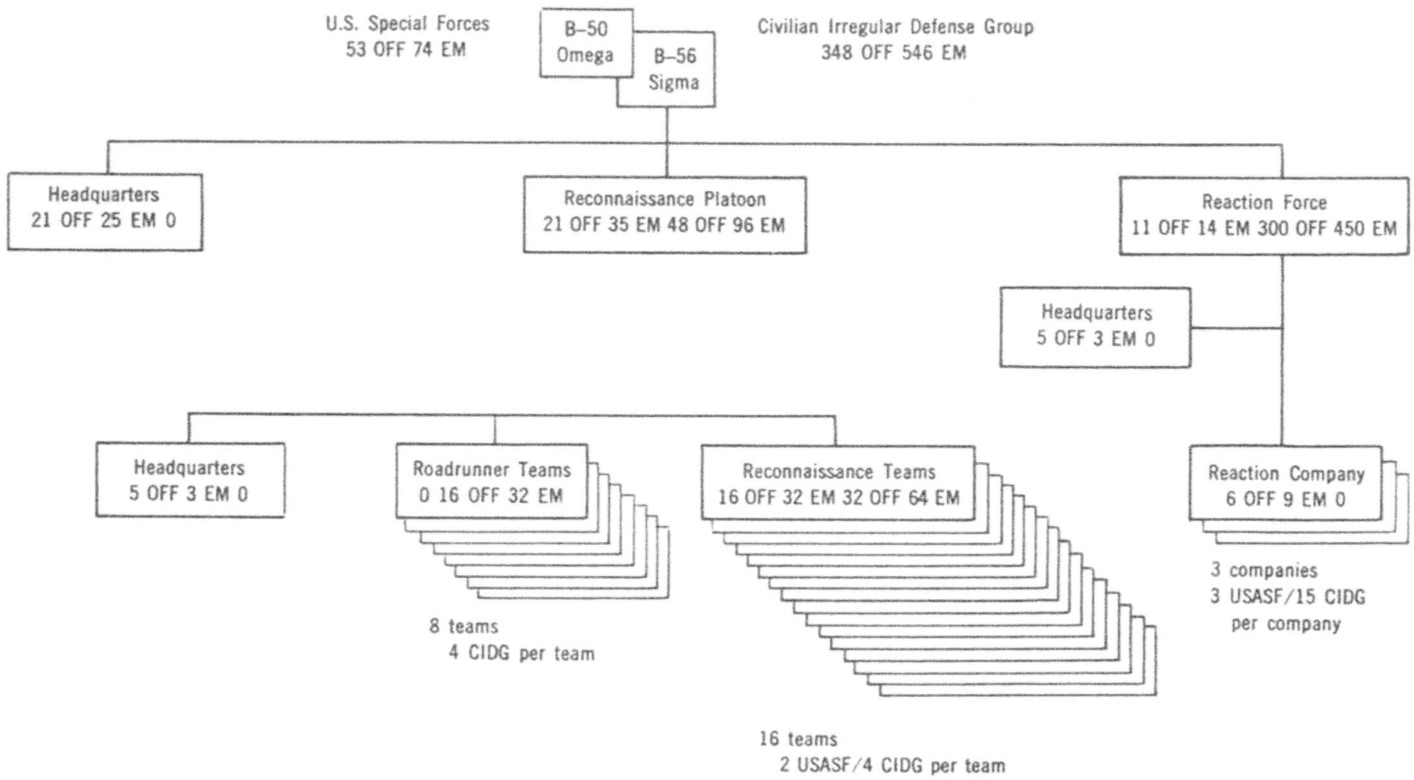

detailed intelligence on enemy movements, routes, and installations, as well as conducting detailed terrain analysis. Backing up the reconnaissance elements of these projects are three MIKE Force reaction companies. These units are employed to exploit small unit contacts, to aid in the extraction of compromised teams, and to perform reconnaissance-in-force missions. Although operational for only nine months, these units have already gathered much valuable information on enemy locations, movements, and lines of communication. They have made numerous sightings of enemy units ranging from squad to battalion size forces. Exploiting these sightings, the teams called in TAC airstrikes on the enemy when feasible. It is noteworthy that since their activation, these units have spent an average of 60% of their time on operations. In addition to gathering much valuable intelligence information, they have accounted for 191 enemy KIA (USASF body count). Probably more important, however, is the psychological impact which these units have on the enemy. The enemy is beginning to realize that he no longer has exclusive dominion over his safe areas. As techniques and training progress, the detachment efforts will be appreciably increased.

Project Delta (B-52) was organized in 1964 under VNSF command with USASF advisors. Since it was used as the basis for the organization of Projects Sigma and Omega, its composition, capabilities, and limitations are essentially the same. The primary difference in organization is the existence of the 91st ARVN Airborne Ranger Battalion as the reaction and reinforcing unit. In view of the dual command relationship, all requests for this unit had to be submitted to JGS and MACV through the respective channels. During the past year this unit has been operating most frequently in I CTZ; however, it has the capability of deploying into any corps when so directed by JGS and MACV. While not as directly responsible to command requirements as Projects Omega and Sigma, this force has made significant contributions to the overall effort. Because of its increased firepower through the employment of the ARVN battalion as its reaction force, it has been able to account for a greater number of enemy kills (194) than the other two units combined. During the past year, it has been deployed in the field, on the average, 55% of the time.

Unit training for unconventional operations consisted of repetitive practical exercises in advanced infantry and special warfare tactics and skills applicable to the environment of mainland Southeast Asia. The training was simplified to the utmost for the benefit of the largely illiterate ethnic and religious minority groups who comprised the forces. The men taken into the unconventional operations forces usually had served previously in CIDG camp strike forces and had therefore been through one or more basic light infantry training cycles. Training for unconventional operations proceeded from this base of knowledge. As a first step, all the men had to qualify for airborne operations. After that the important subjects covered were silent movement; methods of tracking and observation; use of maps and compass; use and care of signaling devices; methods and techniques of infiltration and exfiltration of reconnaissance zones and areas of operations; use and care of special weapons; care and treatment of minor wounds and illnesses; methods of execution of raids and hasty ambushes; defense of bivouac or mission support sites; and procedures for "sterilizing" landing zones and stopover points.

The initial training period covered five to six weeks and was based on a six-day work week and a nine-to-ten-hour work day. Training took place both in base camp and in the field on exercise.

The command and control structure of this type of force was the simplest that could be devised consistent with its organization, missions, capabilities, and support requirements. The mobile strike forces, for instance, were very effective when employed in the roles for which they were trained, that is, reconnaissance and rapid reaction. Unfortunately, when mobile strike forces were attached to U.S. or Vietnam Army units, they were sometimes used as conventional infantry over extended periods of time. This put considerable strain on the strike forces and occasionally resulted in criticism from U.S. commanders of conventional units. Though efforts were made to educate U.S. commanders about the capabilities and limitations of the mobile strike forces, misuse of them remained a continuing problem. Command and control arrangements between special operations forces and higher headquarters varied with the missions and task organization of the forces. For the most sensitive and dangerous missions, command and control were normally exercised directly and jointly from the highest U.S. and Vietnamese headquarters to forces in the field. As missions were progressively less sensitive in nature, command and control was passed to subordinate headquarters.

With regard to air support for special operations, Army aviation attached to the 5th Special Forces Group provided some direct helicopter assault and lift support, and the U.S. Air Force provided airlift for large troop and supply movements. The U.S. and Vietnamese Air Forces rendered other direct aerial support and also provided airborne and other tactical radio-relay links as required. In I Corps U.S. Marine Corps aviation provided direct support in similar ways.

Unconventional operations were an extremely important function of the Special Forces in South Vietnam. With but few exceptions, each operation was carefully planned in advance, and forces were staged from main to forward bases before commitment. The time required to plan an operation and to stage the forces in the area varied with the type and urgency of the mission.

Planning began with a detailed assessment of the designated operational area. All available data on the physical and human geography of the area and on the probable and possible location of enemy forces and facilities within it were collected and analyzed. Relevant data were displayed on maps, and primary and alternative routes of movement for the unconventional operations force were selected and marked. Sources of data for area assessments were maps, intelligence reports, aerial photography, prisoner interrogation reports, and after action reports of friendly forces that had previously operated in the area. If possible, the area assessment planning staff checked the accuracy of its assessment through visual reconnaissance flights over the proposed operational area. Such flights were especially important for the final selection of primary and alternative helicopter landing zones for infiltration and extraction.

When the area assessment was completed, operational orders were issued, and the unconventional operations force was staged if necessary to a forward operating base. Staging involved the transport of men and matériel and was usually accomplished by airlift. If only a few men and a small amount of matériel were to be moved, Army or Marine Corps helicopters and light fixed-wing aircraft were sufficient to accomplish the lift. For larger operations, U.S. Air Force C–7A and C–130 aircraft were employed.

On arrival at the forward operating base, the force was inspected for readiness and the tactical command element presented a "briefback" to the senior officers responsible for operational command, control, and support immediately before commitment. The briefback amounted to a detailed presentation of the operational plan and was designed to insure that every tactical commander, and for small teams every member, knew precisely what his responsibilities were as well as how, when, and why he had to discharge them under the widely differing sets of circumstances that could be encountered during the operation.

For reconnaissance missions, the precise methods and procedures for accomplishing each task associated with the mission were presented in exceptional detail. All were the products of the hard experience of Vietnam. Covered in the briefback were such items as the order and manner of exit from the helicopter that was to introduce the team into the reconnaissance zone, the schedule of movement within the area, procedures for breaking contact with the enemy, and the schedule for and manner of leaving the area.

Much of the success of unconventional operations depended on surprise. In addition to stringent security to safeguard plans, numerous measures were employed to deceive the enemy. Deception was most important at the outset of the operation. In the manner common to Vietnam, the force infiltrated by land, air, or water as befitted the locale, the season, and the size of the force. If, for example, the area of operation was not normally and routinely overflown by friendly aircraft or lacked suitable helicopter landing zones or parachute drop zones, the force might infiltrate overland or perhaps by small watercraft. The helicopter was, however, the usual means of infiltration.

Experience in Vietnam showed that infiltration by helicopter was best accomplished at last light when the pilots could still see well enough to insert the force and have a few minutes to slip away from the landing zone as both force and helicopters were enveloped by protective darkness. Since the enemy was familiar with this method of infiltration, it was necessary to deceive him in regard to the exact point of landing. The helicopters therefore often set down briefly at three or more points in the vicinity of the primary landing zone to create uncertainty in the enemy's mind as to the exact point of insertion. A variation of this technique was also employed when small reconnaissance parties were inserted. A trio of helicopters would fly low in trail formation with sufficient separation to afford the lead helicopter time to touch down momentarily, discharge its reconnaissance team while the other aircraft passed over the landing zone, and rejoin the flight as the last machine in trail.

Communist forces in South Vietnam were very sensitive to these operations and adopted simple but effective countermeasures against infiltration of their refuges. Chief among these was planting long bamboo poles upright in jungle clearings potentially useful as helicopter landing zones, densely covering such clearings with punji stakes, and assigning guards to clearings in the vicinity of their troop units and installations. Clearings studded with bamboo poles were easily recognized by staff planners during aerial reconnaissance of prospective operational areas. Punji stakes in high grass, as well as the presence of guards, were seldom detected beforehand, however, and were often encountered. Casualties resulting from punji stakes, detection by guards, or any overt sign that the enemy had been alerted to the infiltration were causes for immediate evacuation of the party.

Under some circumstances infiltration was best accomplished on foot. Roadrunner and reconnaissance teams were quite easily inserted into a reconnaissance zone from a base camp under cover of darkness, or even during daylight hours if the camp was known to dispatch small patrols in random directions as a matter of routine. If such was not the case, the special reconnaissance team could always leave the base as a part of a larger patrol force and then quietly break away from the force at a preselected time and place.

A similar technique was also employed for a company-size or larger force. After establishing a routine of departing from and returning to the base at random intervals of time and different directions, the force would use an indirect approach to penetrate its operational area. This technique offered a high probability that the enemy would fail to detect the penetration and that enemy agents in the vicinity of the friendly base camp would be unable to report an unusual development in camp operations.

All movement by unconventional operations forces was carefully planned. The survival of small reconnaissance teams depended on each individual team member's knowing and rigidly following a precise route and schedule of movement. The plan might provide for the deliberate and temporary separation and subsequent rendezvous of team members, but it had to provide for rendezvous at precise times and locations if separation occurred under enemy pressure. Mobile strike force movement was planned in less exacting detail but nonetheless followed selected routes unless terrain, vegetation, an engagement with the enemy, or the unexpected appearance of a lucrative target of opportunity justified a change in plan. With more men and heavier firepower, a mobile strike force was better equipped to engage and defeat the enemy.

Stealth was the principal characteristic of movement by unconventional operations forces. Though the enemy might soon become aware of the presence of the men, it was essential that he remain ignorant of their exact location. Movement had to be as silent as possible. Hand and arm signals were used instead of voice commands; voice radio contacts were held to a minimum; weapons and equipment were padded or taped to prevent rattling or metallic sounds when they were brought into contact with rocks or underbrush; and march silence was strictly observed.

The enemy proved quite adept at detecting and tracking such forces even when these precautions were taken. His countermeasures consisted mainly of placing guards at such places as trail junctions and stream crossing points to signal information on the movement of the force by means of a simple code of rifle fire and by having a few trackers follow the force at a safe interval to chart and report on its movements. The enemy also monitored voice radio frequencies normally used by friendly forces for tactical command and control. Feints, ambushes, booby traps, frequent changes in the apparent direction of movement of the force, and strict radio silence were used against the enemy countermeasures.

Suitable sites for bivouac, rest, resupply, and the temporary basing of the force in the field were carefully assessed and plotted before each mission. Sites were designated as primary or alternative, according to the whole plan of movement and the known or estimated adequacy of the cover, availability of water, and defensibility of the terrain. Such operations were aggressive by nature and therefore a force seldom expected to occupy a refuge site for more than half a day. An important exception to this general rule was a temporary base or mission support site for mobile strike operations. From this base a strike force could break down into patrols of squad and platoon size to comb a suspected enemy refuge area. Prolonged use of a mission support site, however, was avoided because it invited enemy attack.

Refuge sites used by the force were carefully policed when the force departed. Enemy trackers were always quick to search these evacuated sites for scraps of intelligence on the strength and intentions of the force, and they frequently dug into garbage pits for such clues. Unconventional operations forces found it useful to booby-trap garbage pits to discourage such probing. Self-destruction devices had, however, to be so employed as to prevent injury to other friendly forces and friendly noncombatants who might occupy the site at some later time.

The nature of the internal defense mission of the U.S. Special Forces in Vietnam required the development of new techniques for resupply in the field. Experience showed that unconventional operations forces, regardless of size, could not carry much more than a five-day supply of food, ammunition, and other necessities. Accordingly, each operational plan provided for resupply at intervals of three to five days at predetermined sites.

In an unconventional warfare role, the forces were expected to live off the land and to replenish at least some of their ammunition and matériel from captured enemy stores and caches of clandestinely acquired items. There was no need to follow such procedures in an internal defense role where resupply was relatively easy, yet the necessity to avoid betraying the location of the force to the enemy remained. Standard airdrops in daylight hours were easily observed by an alert enemy, and night drops in jungled terrain stood a slim chance of being efficiently recovered.

Every plan for unconventional operations had to provide for

SPECIAL FORCES SOLDIER DIRECTS EMERGENCY REMOVAL OF TROOPS

both the routine and emergency removal of individual members or the entire force from the designated area of operations. Planning was necessarily contingent on the innumerable factors that could influence the immediate tactical situation. Under the best of circumstances, and after realization of its objectives, a force could simply walk out or be picked up by aircraft and lifted out of its operational area at a planned point in time and space. The unconventional operations of U.S. Special Forces in Vietnam were so successful through mid-1967 that most were terminated according to plan.

The probability that a force would have to be recovered from an operational area before the planned termination increased in inverse proportion to the size of the force committed. Roadrunner and long-range reconnaissance teams were the most vulnerable to destruction by enemy forces; therefore the command and control element of the force had to be prepared to extract teams from reconnaissance zones at a moment's notice. Ordinarily helicopters lifted out small patrols. If time and circumstances permitted, they would touch down momentarily and recover in standard fashion.

If the friendly force had to be recovered from a position unsuitable for touchdown, a block-and-tackle rig was employed for hasty lift. The typical rig, called a McGuire rig after the Special Forces sergeant who devised it, was a simple rope sling into which a man on the ground could quickly fasten himself in a seated position or, under urgent circumstances, simply affix to his wrist and be plucked from danger in a matter of seconds by the rapid ascent of the hovering aircraft.

Procedures for the evacuation of casualties and for the emergency evacuation of a force larger than a reconnaissance team were more complex. The decision to evacuate or extract posed a difficult problem that required careful weighing of such factors as the mission requirements; the constraints of weather, time, and the position of the force; and the degree of danger which faced the force and the evacuating aircraft. The decision had to be made without delay, and eventualities had to be provided for in the operational plan. It was essential that alternative procedures be explained and rehearsed under simulated conditions in the training that preceded a mission.

BLACKJACK 33, a typical unconventional operation, was carried out between 27 April and 24 May 1967 in III Corps. It was the first operation in which a mobile guerrilla force was employed in conjunction with the long-range reconnaissance capability of a project force—Project Sigma, Detachment B–56. The operation was highly effective; 320 of the enemy were killed. BLACKJACK 33 was commanded by Lieutenant Colonel Clarence T. Hewgley and Captain James Gritz. (*See Appendix F.*)

A basic misconception of the nature of Special Forces operations in Vietnam was created by publicity attendant to the U.S. Army's arrest in July 1969 of Colonel Robert B. Rheault, then commanding the 5th Special Forces Group (Airborne), six officers of an intelligence detachment attached temporarily to the 5th Special Forces Group, and a Special Forces noncommissioned officer. The eight were charged in connection with the alleged murder of a South Vietnamese intelligence agent suspected of being a double agent.

In September Secretary of the Army Stanley R. Resor announced a dismissal of charges because the Central Intelligence Agency had determined that, for security reasons, its personnel could not be made available as witnesses, and without them Mr. Resor deemed that a fair trial was impossible. Since no trial was conducted, the accuracy of the charges and of contemporary newspaper accounts was never ascertained. The public impression created by this case, in which only two of the principals were members of the Special Forces, was misleading and diverted attention from the invaluable contributions of the Special Forces in a wide range of other activities.

One final aspect of the unconventional operations of the 5th Special Forces Group deserves mention—the efforts made by the group to recover prisoners of war held by the enemy in South Vietnam. The operations were directed at liberating any and all prisoners of whatever nationality. Operations with the specific mission of recovering prisoners were mounted and conducted throughout 1966 and 1967. In the fall of 1966 an operation using mobile strike forces was mounted to recover prisoners being held in camps in the U Minh Forest in IV Corps. Although a sharp firefight ensued, no prisoners were liberated. Early in 1967 an operation was conducted in Tay Ninh Province, again by mobile strike forces, to pin down prisoner of war camp locations. Another operation in early 1967 was concentrated in the An Loa Valley in II Corps but no camps were located. In the spring of 1967 mobile guerrilla forces participated in a prisoner recovery operation—part of BLACKJACK 41—in the Seven Mountains region. Also in the spring of 1967, raids on prisoner of war camps in War Zone C of III Corps were staged out of the CIDG camp at Cau Song Be. Project Sigma forces together with mobile strike forces participated in these operations. While several camps were overrun, they were found to be deserted. Operations to recover prisoners of war were a constant objective, even though they were unsuccessful. Despite the cost in men, intelligence effort, and operational assets, these operations were mounted whenever and wherever possible, but the Viet Cong used the tactic of constantly moving prisoners of war from one place to another in order to foil external liberation efforts and internal escape plots.

The significance of the unconventional operations conducted by the 5th Special Forces Group is that the two types of operations—counterinsurgency and unconventional—could be carried out successfully and simultaneously. These unconventional operations were a source of pride to the Special Forces soldier; in fact most of the troops were originally attracted to the Special Forces by the nature of these operations. A testimony to the flexible organization of the group was the ability of the logistic sections to mount and support such a variety of operations.

PART THREE

THE LATTER YEARS: 1968–1971

CHAPTER VII

The Vietnamese Army Absorbs the CIDG Program

The major task of the Special Forces during the last two and a half years of the 5th Group's stay in South Vietnam was to complete the turnover of the Civilian Irregular Defense Group program to the Vietnamese. The concept of Vietnamization, which became the focal point for all U.S. strategy in the period 1968–1969, was not new to the 5th Special Forces Group (Airborne). The Vietnamese Special Forces, however, had been slow to develop soldiers of high professional skill, partly because the introduction of large U.S. regular military forces had made the war in Vietnam a more conventional conflict. Nevertheless, command emphasis from the 5th Special Forces Group continued to be placed strongly on the importance of delegating responsibility to the Vietnamese. The goal of conventional forces was the conventional one of winning the war. For Special Forces, however, the goal was to help the South Vietnamese win what was really their war, and that goal was never forgotten. A victory or defeat was a victory or defeat for the people of Vietnam, represented by the indigenous Civilian Irregular Defense Group troops.

The U.S. Special Forces troops were in combat right up until the day they left. New camps continued to be built and old ones were fortified and strengthened in preparation for the Vietnamese take-over.

Plans for the transfer of the Civilian Irregular Defense Group program to the Republic of Vietnam were considered as early as 1964. The increased tempo of the war after the 1968 *Tet* offensive did not permit the regular armed forces of Vietnam to take over border surveillance until 1970. By 1969 it was apparent, however, to the Vietnamese Joint General Staff and the Military Assistance Command, Vietnam, that the Vietnam Army had reached a level of competence that would enable it to take over the additional missions of border surveillance and interdiction.

Although Special Forces CIDG camps were regularly converted to Regional Forces status or closed down when an area became pacified, the phase-down directed in the early part of 1970 envisioned the discontinuance of the entire CIDG program. A combined Joint General Staff–Military Assistance Command planning committee was convened on 20 March 1970 for the purpose of terminating the CIDG program in a smooth and orderly manner. The committee recommended that all remaining CIDG camps be converted to the Vietnam Army between the months of August and December 1970. This conversion consisted of absorbing the CIDG men into the Vietnam Army where they would become government soldiers instead of civilian irregulars. (*Table 8*) A border control system using Vietnam Ranger battalions was decided upon as the appropriate successor to the Civilian Irregular Defense Group. Vietnam Special Forces were to assist the Rangers in a recruiting drive aimed at converting the outgoing civilian irregulars into Rangers. Toward that end, the Vietnam Special Forces participated, along with the U.S. Special Forces, in a motivation and indoctrination program that explained the benefits of conversion to the CIDG troops. Figures on the number of civilian irregulars who volunteered varied from camp to camp, but the majority chose to convert to Rangers. (*See Appendix C.*)

During 1970 combat continued, but at a somewhat reduced tempo. The incursion into Cambodia in the spring of 1970, in which the CIDG participated, had significantly weakened the enemy in Vietnam. Pressure on the camps, especially in the III Corps Tactical Zone, decreased noticeably after the Cambodian operation.

TABLE 8—CONVERSION OF U.S. SPECIAL FORCES CIDG CAMPS TO VIETNAM ARMY RANGER CAMPS, 27 AUGUST 1970–15 JANUARY 1971

Camp	Conversion Date	Ranger Battalion Designation	Number of Men
I Corps Tactical Zone			
Mai Loc (A-101)	27 August 1970 (closed)		
Ha Thanh (A-104)	31 August 1970	87th	415
Tra Bong (A-107)	31 August 1970	61st	486
Minh Long (A-108)	30 September 1970	68th	386
Ba To (A-106)	30 September 1970	69th	400
Gia Vuc (A-107)	30 September 1970	70th	460
Tien Phuoc (A-102)	31 October 1970	77th	422
Nong Son (A-105)	31 October 1970	78th	392
Chu Lai (B-11)	31 October 1970 (closed)		
Da Nang (Company C)	1 November 1970 (closed)		
Thuong Duc (A-109)	15 November 1970	79th	396

TABLE 8—CONVERSION OF U.S. SPECIAL FORCES CIDG CAMPS TO VIETNAM ARMY RANGER CAMPS, 27 AUGUST 1970–15 JANUARY 1971 —Continued

Camp	Conversion Date	Ranger Battalion Designation	Number of Men
II Corps Tactical Zone			
Polei Kleng (A-241)	31 August 1970	62d	403
Plei Mrong (A-113)	31 August 1970	63d	443
Tieu Atar (A-231)	30 September 1970	71st	414
Trang Phuc (A-233)	30 September 1970	72d	399
Plei Djereng (A-251)	31 October 1970	80th	479
Duc Co (A-253)	31 October 1970	81st	457
Plei Me (A-255)	31 October 1970	82d	464
Bu Prang (A-236)	30 November 1970	89th	377
Dak Pek (A-242)	30 November 1970	88th	298
Dak Soang (A-245)	30 November 1970	90th	431
Kontum (B-24)	30 November 1970 (closed)		
Ban Me Thuot (B-23)	15 December 1970 (closed)		
Ben Het (A-244)	31 December 1970	95th	430
Duc Lap (A-239)	31 December 1970	96th	400
Pleiku (Company B)	15 January 1971 (closed)		
III Corps Tactical Zone			
Tra Cu (A-316)	31 August 1970	64th	334
Ben Soi (A-120)	31 August 1970	91st	386
Thien Ngon (A-323)	30 September 1970	73d	333
Loc Ninh (A-331)	30 September 1970	74th	358
Duc Hue (A-325)	31 October 1970	83d	253
Katum (A-375)	31 October 1970	84th	369
Tay Ninh (B-32)	31 October 1970 (closed)		
Hon Quan (B-33)	30 November 1970 (closed)		
Tong Le Chon (A-334)	30 November 1970	92d	318
Trang Sup (A-301)	30 November 1970	65th	427
Bu Dop (A-341)	31 December 1970	97th	300
Bien Hoa (Company A) ...	1 January 1971 (closed)		
IV Corps Tactical Zone			
To Chau (A-442)	31 August 1970	66th	379
Thanh Tri (A-414)	31 August 1970	67th	315
Tuyen Nhon (A-415)	30 September 1970	75th	302
Cai Cai (A-431)	30 September 1970	76th	398
Binh Thanh Thon (A-413)	31 October 1970	86th	332
Chi Lang (A-432)	31 October 1970	85th	210
Moc Hoa (B-41)	31 October 1970 (closed)		
Ba Xoai (A-421)	30 November 1970	94th	408
Vinh Gia (A-449)	30 November 1970	93d	460
Can Tho (Company D) ...	1 December 1970 (closed)		
Chi Lang (B-43)	1 December 1970 (closed)		

Summary by Corps of Combat Activity During 1970

The North Vietnamese Army and the Viet Cong continued to be active in I Corps during the year. Company C, 5th Special Forces Group, was charged with Special Forces operations in I Corps. The camp at Tien Phuoc achieved one of the highest kill ratios of any camp in the zone in 1970. For a period of two to three months, the camp averaged 50 to 60 of the enemy killed each month and itself suffered few casualties. The camp at Mai Loc received an early morning sapper attack in which most of the camp structures were destroyed before the CIDG troops succeeded in driving the Viet Cong out. Shortly after the attack on Mai Loc, the camp at Thuong Duc was taken under siege by the enemy, who used mortar and rocket barrages. The siege lasted sixty days, but the camp held out. Again, in October, Thuong Duc came under attack, but the camp defenders seized the initiative and in a three-day period killed 74 of the enemy by small arms alone. Over a seven-day period, three heavy battles resulted in a final total of 150 of the enemy killed.

Siege is the key word to describe the combat activity in the II Corps Tactical Zone in 1970. The Special Forces activity in II Corps was controlled by Company B. Bu Prang, which had undergone a 45-day siege at the end of 1969, was rebuilt completely underground. *(See Appendix D.)* Dak Seang was taken under siege at 0645 on 1 April, and when it became clear that the enemy was making a determined attempt to destroy the camp reinforcements were sent in. Mobile strike forces and Vietnamese Ranger battalions came to the aid of the camp and helped to inflict heavy casualties on the enemy forces—the 28th North Vietnamese Army Regiment, the 40th North Vietnamese Artillery, and elements of the 60th North Vietnamese Army Regiment. Twelve days after the beginning of the attack on Dak Seang, the enemy turned his siege tactics on Camp Dak Pek, attacking with mortars before dawn, and following with a sapper attack. While the camp was almost completely destroyed, enemy losses were extremely high. A lull set in after the thrusts into Cambodia, and handling refugees became the major task in II Corps camps. Refugee villages were established near Bu Prang and Duc Lap.

The Cambodian operation in which Company A, 5th Special Forces Group, participated was the high point in combat operations in III Corps in 1970. Civilian Irregular Defense Group companies from the camps at Duc Hue and Tra Cu assaulted a Viet Cong training area in Cambodia and discovered caches that provided over one-third of all the crew-served weapons captured during the May offensive. The Civilian Irregular Defense Group troops not only captured equipment, but also killed eighty of the enemy. Earlier in 1970 the III Corps CIDG Mobile Strike Force distinguished itself by capturing an enemy cache of record size. The action took place in War Zone D near Rang Rang, an area which traditionally had been an enemy stronghold. The strike force picked up 450 SKS carbines, 1,034 82-mm. mortar rounds, 130 122-mm. rockets, and almost 200 tons of rifle and mortar ammunition. The camps at Katum, Tong Le Chon, and Bu Dop were all subjected to heavy mortar attacks, with Katum, the heaviest hit, receiving for one period close to 300 rounds a day. After the Cambodian operations, as in II Corps, things quieted down.

Combat operations in IV Corps continued to be centered around the special techniques the 5th Group had developed for using the waterways in the Mekong Delta to wage war. The Delta Company Strike Force made heavy use of sampans and airboats in its operations. The Seven Mountains region was again the scene of heavy fighting when strike force troops, along with strike forces from Camps Ba Xoai and Vinh Gia, moved in to clear the area. Ba Xoai and the B detachment at Chi Lang both repulsed enemy attacks on those camps. As a whole, IV Corps operations indicated that in the delta especially the Vietnamese were fully capable of running the war on their own.

Civic action under the Special Forces Civilian Irregular Defense Group program was one of the most important efforts of the war. Whether of immediate or long-range value, the missions demonstrate the commitment the U.S. Special Forces had to the people of Vietnam.

A summary of the civic action missions of the 5th Special Forces Group in the period 1964–1970 shows that the group set up 49,902 economic aid projects, 34,334 educational projects, 35,468 welfare projects, and 10,959 medical projects; furnished 14,934 transportation facilities; supported 479,568 refugees; dug 6,436 wells and repaired 1,949 kilometers of road; established 129 churches, 272 markets, 110 hospitals, and 398 dispensaries; and built 1,003 classrooms and 670 bridges.

The Stand-Down

In military usage, the order to stand down is employed to discontinue a condition of battle readiness or some varying degree of alertness to the possibility of combat. The stand-down period normally is very short and is characterized by rapid relief of the unit from the combat environment, frequently under the protection of supporting, adjacent, or relieving units. When the order to stand down was received by the 5th Special Forces Group, few if any of the usual characteristics of the operation applied or were in evidence. Seldom does a unit being relieved or replaced have to plan, train, and provide for its replacement forces. This was precisely what faced the U.S. Special Forces as they prepared to leave Vietnam after ten years of fighting.

By 1 June 1970, the number of Civilian Irregular Defense Group camps in Vietnam had been reduced to thirty-eight, either by conversion to Regional Forces status or by closure. The Vietnamese Joint General Staff and the Military Assistance Command, Vietnam, staff then decided to convert the remaining camps to Vietnamese Army Ranger camps, with a target date of 31 December 1970. (*See Appendix A.*) Progressive, concurrent conversion cycles were initiated, with the major criteria being the state of security around each camp and seasonal weather. Camps in relatively secure areas that could be supplied easily during the rainy season were converted first. Camps in less secure areas were scheduled for later conversion so that more time and resources could be applied to increasing the combat readiness of these camps. One camp, Mai Loc, was closed entirely, so that the final number of CIDG camps converted to Vietnam Army Rangers was thirty-seven.

There were many significant circumstances attending the conversion process. The 5th Special Forces Group was working toward a date for redeployment to the continental United States of no later than 31 March 1971. It was also supporting the thirty-seven camps in widely separated and remote locations in all four corps tactical zones of South Vietnam. As the transfer of CIDG camps and assets to the Vietnam Army Ranger Command proceeded, the 5th Special Forces Group was being reduced in size while the group administrative, operational, and logistical phase-down problems increased. U.S. Army advisers to the Vietnam Army Ranger elements at converted camps were not available; three Special Forces soldiers were therefore required to remain at each location after conversion until the U.S. Army advisers arrived. The status of CIDG paramilitary members had to be changed to that of regular Vietnamese Army soldiers. Inventories of supplies and transfers of real estate had to be accomplished under both U.S. and Vietnam laws, statutes, and regulations. And, finally, the complex operations of inactivation and redeployment of 5th Special Forces Group units and individuals had to be undertaken simultaneously with the conversion schedule.

The major considerations in the stand-down are related in

detail in the final after action report of the group, and include administration, operations and intelligence, logistics, and real estate. Administratively, two basic problems that confronted the group were the transfer of CIDG men from the paramilitary status in which they were a responsibility of the 5th Special Forces Group to regular Vietnam Army soldier status, and the maintenance of control over personnel actions affecting the U.S. Army Special Forces soldier during the stand-down and redeployment phases.

The specific plan for absorbing members of the CIDG program in the thirty-seven camps transferred to the Vietnam Army Ranger border defense battalions was developed, directed, and carried out by the Vietnamese Joint General Staff. The 5th Group interests in the transfer were in payroll procedures, severance pay, bonuses, and other fiscal arrangements made with the paramilitary civilians. The stand-down of CIDG men was accomplished with relative ease and efficiency.

The U.S. Army Special Forces was chiefly concerned with maintaining unit strengths at prescribed levels and providing adequate troop services during the redeployment process. Generally, U.S. Special Forces men who had spent less than ten months in Vietnam—some 1,200, or 60 percent of the group strength—were reassigned to other U.S. Army, Vietnam, units. The remainder returned to the continental United States. Legal matters, including pending court-martial cases, were solved or, as a last resort, referred to Headquarters, U.S. Army, Vietnam. The civilian labor force was abolished but substantial efforts were made to find positions for the affected employees. Records and files were closed out in accordance with standing regulations and the projected timetable. An effective use was made of the information office to keep all group troops continuously informed of the progress of redeployment.

Conversion of the U.S. Special Forces CIDG program to the Vietnamese Army Ranger program was actually an expansion of a long-term transfer program begun in 1964. The principal objective of both efforts was to help the Vietnamese armed forces to conduct the war with less and less assistance from the United States. During the stand-down period every effort was made to raise the combat readiness of the thirty-seven remaining CIDG camps to the highest efficiency. Concurrently, a concerted effort was made to assimilate the Montagnard and other minority ethnic groups from remote areas into the Vietnam Army. The Vietnamese Special Forces and the 5th Special Forces Group staffs developed jointly a program designed to continue operational missions in CIDG camps; process CIDG members administratively and medically; prepare U.S. Military Assistance Command, Vietnam, advisers for camp missions; transfer logistical support; reorganize CIDG units into Ranger battalions; and assimilate CIDG leaders into the Vietnam Army ranks.

The conversion process proceeded successfully, partly because the Vietnamese Special Forces camp commanders stayed in place and automatically became Ranger battalion commanders. Their familiarity with the troops, the camp area, and the tactical area of operations was invaluable. The Military Assistance Command, Vietnam, advisers did not arrive for duty until some seventeen camps had been converted. The fact that many of the advisers were former Special Forces men familiar with the camps minimized the problem.

As a result of the close co-ordination between U.S. and Vietnamese Special Forces, the Vietnam Army Ranger Command was strengthened by the addition of thirty-seven light infantry battalions. Of the possible 17,057 troop spaces scheduled for conversion, 14,534 CIDG troops actually became members of the Ranger command. A significant benefit that accrued to the minority ethnic groups involved was better treatment by the government of Vietnam. For their allegiance, as expressed by their willingness to join the Vietnam Army units, the government provided legal birth and marriage certificates as well as medical benefits and disability pay for injuries received in military action. This was the first time that the minority groups, and particularly the Montagnards, were given full status as citizens of the Republic of Vietnam.

At the beginning of the stand-down period, the troops needed for logistic support totaled almost 30,000, of which 2,000 were U.S. Special Forces and the remainder CIDG members, stationed at some fifty-four locations throughout the country.

Because of the variety of functions and the many regulations involved, a carefully developed plan of interlocking supply and maintenance actions began. (*See Appendix* G.) There was enough time allowed for an efficient close-out of supplies, equipment, and matériel. As could be expected, many minor problems arose which were either local, relatively inconsequential, easily solved, or all three.

There are a number of lessons to be learned from the close-out of the logistic program. In the future it would be wise to incorporate data processing into group logistic activities and the automated system should be adaptable down to forward supply point (company) level wherever possible. Documentation for logistic activities must also become automated and streamlined. In any stand-down operation, security of supplies must be increased geometrically. Pilferage and lost or misplaced items are the inevitable consequences of the turnover of a supply system, and safeguard measures must be employed promptly and enforced rigidly. Collection areas are particularly vulnerable and constant and scrupulous attention to security procedures is mandatory. Obviously any overstockage of items reduces efficiency.

The real estate problems of the 5th Special Forces during stand-down were many and varied. First, the group had extensive holdings throughout Vietnam because of its mission; some of the camps dated back to 1962. At the time of acquisition, most of the regulations dealing with the administrative aspects of returning real estate had not been written; therefore, many of the documents deemed necessary to accomplish the return had never been collected. Second, the U.S. Army fiscal and logistic regulations governing U.S. Special Forces operations were ambiguous when dealing with real estate. Third, many Special Forces headquarters sites were in highly desirable locations, creating a competition between U.S. and Vietnam Army units for occupancy. Fourth, project planning during the phase-down operated under several restraints. Numerous changes in turnover dates occurred. Also, stipulations as to maximum conditions caused engineer missions to continue up to the actual date of turnover. The only complex turnover parcel concerned the group headquarters location at Nha Trang. Despite numerous false starts, this bulky parcel was turned over within the scheduled time limit.

The problems encountered in the engineer and real estate transfer were the results of circumstances so peculiar to this unit and locale that they are perhaps applicable to no other, certainly not in the foreseeable future.

The participation of the 5th Special Forces Group in the Civilian Irregular Defense Group program, like the program itself, ended on 31 December 1970. The program was in many ways a chronicle of the larger war. Developed in response to the needs of the Vietnamese Army, the government, and the Free World Military Assistance Forces, the program and the 5th Special Forces Group displayed an organizational flexibility and competence in the field that is rare in the annals of modern warfare. The U.S. Army Special Forces came home from Vietnam confident that the men they had advised, trained, and led would be able to carry on the struggle bravely and well.

CHAPTER VIII

The Balance Sheet

An elite group has always appeared within the Army during every war in which the United States has been engaged. The Minutemen in the Revolution, the Cavalry in the Civil War, the Rough Riders in Cuba, the Lafayette Escadrille in World War I, the Rangers in World War II, and the Helicopter Pioneers in Korea—always some group has captured the imagination of the American public and has embodied the national ideals of the American fighting man.

As surely as such groups arose, there arose also the grievances of the normally conservative military men who rejected whatever was distinctive or different or special. The conservative approach to military matters is, of course, by and large the safest, most effective, and most practical. It is in the American character, however, to attack problems vigorously, to attempt rapid and complete solutions, and to accomplish the business at hand with a certain amount of independent daring and courage. Thus, the emergence of Army units combining these characteristics is not unusual but is the historical pattern. Future planners would do well not only to recognize this American military phenomenon but also to capitalize on it.

In the conduct of conservative military affairs, revisions of current military modes are frequently resisted with missionary zeal and emotional fervor simply because they mean change, they are different. In the complexities of handling national defense matters, a defender of the *status quo* can find many reasons for not doing something. If a new military program or unit is being developed in order to meet new needs, new threats, or new tactics, consideration should be given to the use of elite U.S. Army units, despite the customary resistance to change or elitism usually found in conservative establishments.

The U.S. Army Special Forces had the continuous and unswerving support of each commander of the U.S. Military Assistance Command, Vietnam. Generals Harkins, Westmoreland, and Abrams recognized the value of the Special Forces operations and consistently provided the unit with maximum support, direction, and guidance. Because Special Forces was a unique organization with many talents and demonstrated capabilities, each commander had somewhat different ideas on how to use it. Regardless of the employment, however, each commander was completely receptive to new tactics and techniques, new plans and programs, and new operational possibilities. Operational requests, personnel requisitions, and administrative and logistical demands were promptly and carefully attended to and authorized whenever possible. Whatever shortfall in Special Forces operations may have occurred, it never came as a result of lack of support from the head of the Military Assistance Command.

One single statutory action that proved most beneficial to the Special Forces was the approval in September 1963 of an Army regulation which dealt with the administrative, logistical, and financial support for paramilitary forces and provided the means by which such support could be obtained, managed, controlled, and accounted for. Before that date support arrangements were accomplished on an *ad hoc* basis, leaving no firm, acceptable method for accomplishing these requirements on an approved, departmental, permanent basis. The publication of this regulation for the first time placed Department of the Army support for paramilitary activities on a sound, respectable, businesslike footing.

Support by the U.S. Air Force in the Republic of Vietnam was superb. The tactical air force and airlift command elements performed outstanding feats in support of Special Forces. For example, airlift for the first of the three combat parachute assaults conducted by the Special Forces in South Vietnam consisted of nine C–130 aircraft. These planes were assembled, rigged, operationally prepared, spotted, and ready for take-off within a few hours after the approval of the operation was given. The first aircraft crossed the intended drop zone exactly on the minute prescribed. In October 1966 tactical aircraft, hastily scrambled, provided the firepower to rescue a sizable contingent of Special Forces in the Plei Trap Valley. Without these fighters, the force stood to receive staggering casualties. Tactical aircraft provided instant response to missions generated by the mobile guerrilla forces, including resupply of vital necessities. Airlift command was largely responsible for the movement each month of 17,000,000 pounds of supplies in 500-pound lots to Special Forces camps throughout Vietnam. The armed C–47 gunship was a tremendous help to camps under attack and accounted for the continued existence of camps many miles removed from immediate relief forces or firepower.

The U.S. Navy contributed significantly to the successful opera-

VIETNAMESE SPECIAL FORCES MOVE UP RIVER ON A MISSION

tions of the 5th Special Forces Group, and this assistance took many forms. For instance, the Navy provided its entire inventory of patrol air cushion vehicles in the theater during the monsoon operations in IV Corps in 1966–1967. These craft, together with the Special Forces airboats and motorized sampans, raised havoc with enemy forces slowed down or stopped by the floodwaters in the Mekong Delta. Navy personnel also acted as instructors for the irregulars piloting the airboats and sampans. Navy river patrol craft worked harmoniously and successfully in joint land operations with troops of the Civilian Irregular Defense Group and the Special Forces advisers. Navy SEAL (sea, air, land) teams conducted joint training and exercises with scuba teams from the Special Forces.

The U.S. Marine Corps, especially in the 1962–1964 period, developed an outstanding relationship with the Special Forces. The Special Forces camps in the I Corps area literally lived or died depending on Marine helicopter support in those early days, when supply runs were made into the most rugged areas. Marine helicopters evacuated the survivors of Camp A Shau in early 1966. Joint operations using Marine reconnaissance units with civilian irregular and Special Forces units were most successful, as were certain innovative tactics devised together, such as airlifting

105-mm. howitzers to predetermined meeting sites to attack enemy units preparing to assault outgunned friendly forces. In May 1967 civilian irregulars, Special Forces troops, and U.S. marines fought side by side in defense of the camp at Con Thien, administering a blisteringdefeat to the North Vietnamese Army.

The instances of co-operation and mutual support listed above are very few and do not reveal tbe deep confidence each service had in the other. Rather, they are random examples of common effort, intended to emphasize the truism that service rivalries diminish in inverse proportion to the nearness of the firing line.

In Vietnam there were certain factors that operated against the efforts of the U.S. Special Forces, and over which the Special Forces had little or no control. For the first time in its history the United States found itself waging a military and a political contest simultaneously. The Korean War was limited, as were the numerous incidents between 1945 and 1960. The new factor encountered in the Vietnam conflict was the departure from the sequential "military-then-political" actions of previous wars, in which the military effort was primary and foremost. Decisions to be made were evaluated principally in terms of military consequences, with political implications incorporated as part of a long-range integration of effort. As territory was rolled up, military government forces followed, to be supplanted later by civilian agencies restoring civilian control and development. In Vietnam military decisions were viewed in terms of the political consequences they might have, a situation to which the average military professional was unaccustomed. The usual primary military objectives of "closing with the enemy and defeating him" were limited by political decisions. The immediate impact on the military unit often took the form of misunderstanding, aborted tactical plans, and communication gaps.

There was a lack of understanding throughout all ranks on the nature of insurgent wars and of that in Vietnam in particular. Most U.S. Army schools had failed to incorporate many of the lessons learned in the Korean War. The march and countermarch across the European plain were still the staples for instruction well into the 1960s. Reports from Vietnam that the enemy was a mighty jungle fighter of indomitable prowess, spurred on and nurtured by the knowledge that his political convictions were right, caused the military service schools to juggle hastily the instructional units in the curriculum to accommodate this type of foe. Despite these efforts, the elemental lessons of infiltration, scouting and patrolling, reconnaissance, ambush tactics, night fighting, and unorthodoxy in tactics and logistics had to be learned and relearned on the ground in Vietnam. The twelve-month tour of duty operated against any one commander's accumulating very much experience or passing it on to his successors. The experience of the Special Forces ultimately proved that the night and the jungle belonged to the fighter who could use them best.

The fundamental communications gap stemmed in great part from the education gap. The cliché that the American soldier is the best informed soldier in the world was often repeated but it was sometimes dubious whether he was informed at all. Certainly, in terms of proportionate time, very little effort was made to explain to him Vietnamese or Oriental culture and customs; had the average American soldier been better informed, many actions of the Vietnamese would have been at least understandable, if not palatable to him. So acute was the lack of information that in 1962 special courses on countering insurgencies were hurriedly devised for senior officers, but the course content was a long time getting down to the individual foot soldier. The advice of foreign experts in insurgency and counterinsurgency was sought and followed—even though their expertise, for the most part, had been acquired in different locales under totally different ground rules. As a result the random and usually irrelevent courses of action that were taken had little or no bearing on the Vietnam struggle.

The lack of adequate preparation for the Vietnam War within the active Army took many forms. Not only were the political and sociological aspects of the war given less than full attention, but also the related areas of language training, civil affairs (or civic action, as it became known), psychological operations, and interdepartmental co-ordination received little emphasis. The personnel actions which bothered the Special Forces members most were the complete and continual disregard by departmental personnel officials of the comparative combat responsibilities of Special Forces people. Because the table of equipment for a Special Forces group specified the position of a Special Forces lieutenant colonel as the commander of a Special Forces company, no amount of correspondence ever really convinced the personnel managers at Department of the Army that this position was really the equivalent of that of a battalion commander in terms of combat responsibility.

Command and control rules, procedures, and adjustments suffered because of the lack of understanding of the nature of the war, the lack of education in fighting it on a daily basis, the lack of communication throughout the chain of command, and the inbred convictions acquired during combat operations in World War II and

Korea. The lack of preparation for this war had certain effects. For one thing, it led to a preoccupation with statistics. In many instances the success or failure of an operation was validated by the statistical considerations attending it. The usual method of determining the efficacy of psychological operations, for example, was by counting the number of leaflets dispensed or the number of loudspeaker broadcasts made. Often preconceived operational methodologies were a handicap. Senior U.S. Army commanders arrived with combat methods for the conduct of operations firmly in mind. That the methods did not fit the times or the struggle did not keep the commanders from using them. It often took a substantial period of time to educate such commanders to the facts of life. For example, they were slow to learn that the Vietnamese troops were allies, not subordinates; that the CIDG forces were indeed civilians and irregulars, and CIDG companies did not equate in terms of numbers, firepower, or training with U.S. Army infantry companies; and that all troops of whatever national origin in a given tactical area of responsibility did not automatically come under the command of the U.S. area commander.

There was, too, a certain lack of imagination in the development of new tactics. Deviations from current doctrine, however outdated that doctrine might be, were not systematically sought or encouraged. Unusual formations such as the mobile guerrilla forces and the clandestine resupply methods were accepted because of their demonstrated success. Long-range reconnaissance units, such as Projects Delta, Sigma, and Omega, were welcomed because of their exceptional record of performance, yet reconnaissance teams in major U.S. combat units continued to scout and patrol only as far out as the organic weapons of their units could cover them. The search for combat intelligence still followed the stereotyped pattern of seeking and reporting information, analyzing it so that it became useful intelligence, then seeking more of the same; seldom was a unit concerned with information about the interior control organization—the so-called infrastructure—of the enemy. Fighting camps, floating camps, and waterborne operations in conjunction with helicopters were accepted more as oddities than as adaptations to particular conditions. Night operations were the exception in most units, though as early as October 1966 all irregular operations supervised by the Special Forces began and, where practicable, ended during the hours of darkness.

A sound principle of war deals with the chain of command. This principle holds that orders are best carried out and control and discipline are best maintained by making each level of authority aware of its responsibility for carrying out a mission. The ubiquitous helicopter damaged the chain severely since the temptation to deal with subordinates several layers down was too great to resist. Indeed, the war became known as the "small unit commander's war," quarterbacked by a senior commander circling overhead. With a platoon leader, for example, getting precise instruction from a division commander, the teamwork and leadership development between the platoon leader, the company commander, the battalion commander, and the brigade commander were bound to be disrupted.

Certain factors were controlled by the 5th Special Forces Group in Vietnam and contributed positively to the success of the group effort. The table of organization and equipment remained flexible. Personnel, organizational detachments, equipment, and administrative arrangements could be easily adjusted within the personnel or fiscal ceilings imposed. The Special Forces had many military occupational specialties in its tables. Additional units not of the Special Forces but with similar occupational roles, such as medical or engineer, were easily added, controlled, and supervised. The fragmented, independent operation of the group facilitated *ad hoc* attachments where and when required. The volunteer status of members of the Special Forces effectively weeded out many unqualified men.

Training programs within the Special Forces were of long term, forty-four to sixty-two weeks. Such training combined with follow-up training in a secondary specialty produced soldiers of high professional standing.

The requirement that a man volunteer for both parachute and Special Forces training, the high training standards, the premium on independence and reliability, the emphasis on team loyalty and dedication, and the development of a sense of belonging to an exceptional unit tended to produce the most professional and most capable noncommissioned officers in the U.S. Army. The record of combat decorations, repeated tours in Vietnam, combat efficiency, and manifest pride in the organization reflect this professionalism.

Command of the 5th Special Forces group was placed in the hands of Special Forces officers. In the early days this empirical requirement was not a prerequisite for assignment. Beginning in 1966, the normal policy became a succession of assignments, starting with command of the 1st Special Forces Group in Okinawa and concurrent orientation tours in Vietnam; thereafter, the commanding officer of the 1st Special Forces Group succeeded to command of the 5th Special Forces Group.

TRAINING JUMP

There were also present within the structure of the 5th Special Forces Group in Vietnam certain factors that worked against the efficiency of the group.

In the early years, their role as advisers rather than operators was not made clear to most Special Forces troops. In their desire to accomplish positive gains and as a result of their concern for the welfare of the irregulars, attachments were formed between the Americans and the irregulars which adversely affected the Special Forces effort. The Vietnamese Special Forces initially resented the "big brother" role assumed by many U.S. Special Forces troops; the irregulars, who had relatively little empathy with the Vietnamese, assumed that the U.S. Special Forces would stand with the irregulars against the Vietnamese. These misplaced assumptions were partly responsible for the Montagnard uprising in September 1964. In time both the Americans and the irregulars came to understand and respect the Special Forces advisory role.

It is usually accepted by the military that U.S. Special Forces detachments were more successful in advising local governments than were other U.S. advisory elements. When the full circumstances were known, it was seen that much of the success of the detachments stemmed from their access to "more supplies, more quickly obtained." Detachments doing double duty as advisers for the Vietnamese Special Forces and advisers to local civil government agencies did either job well but seldom both successfully.

The buildup in South Vietnam from 1965 to 1967 outstripped the capability of the continental United States replacement system to furnish trained men. Standards dropped—in the field of communications, for example, the ability to perform at a high rate of speed in continuous wave communications declined—and it became necessary to give each new arrival an examination in his specialty before sending him out to a camp. This additional burden slowed down the operational pace because men had to be diverted to conduct the tests. Lowered standards invited less competent men who could not stand the rigors of the independent, isolated, and perilous life of a camp. From 1966 to 1967 a study was undertaken by the Walter Reed psychiatric unit in conjunction with the 5th Special Forces Group to determine whether it was possible to devise a test or examination that could predict which Special Forces soldiers were likely to break down in a camp environment. After a year of intense and careful study, it was concluded that the best indicator of who would or would not succeed as a positive member of a detachment in the field was the judgment of the senior Special Forces noncommissioned officer who trained or supervised the man. Frequently a Special Forces man would be dropped from the force as not suitable, but by one device or another would regain a position in the Special Forces.

The practice of placing the Civilian Irregular Defense Group camp forces, together with the U.S. and the Vietnam Special Forces detachments, under operational control of Vietnamese and U.S. military commanders was a throwback to conventional lines of command and control. The reasons for such control lines are obvious, but they proved less successful than when the group commander controlled all his men. Aside from the technical, communications, and operational reasons that could be advanced, the average Special Forces man in a camp had better morale and *esprit* when under group command.

Credits

The U.S. Army Special Forces made significant progress in many areas during its term of service in the Republic of Vietnam. As a tactical combat unit, the U.S. Special Forces dealt with the Vietnamese Special Forces in both an advisory and an operational apacity. This primary relationship did not impair or preclude other working relationships with U.S. armed forces, Free World Military Assistance Forces, U.S. government agencies, or Vietnamese government agencies.

Positive contributions were made by the Special Forces over he years to the American national effort to defeat the North Vietnamese Army and the Viet Cong. These contributions were made not only to the immediate operations of the war, but also to the development of the Special Forces as a general purpose unit within he U.S. Army troop structure. As a result, any doubts about the value or practicality of having this type of unit in the permanent J.S. Army force structure were removed.

The record of service performed in the past becomes doubly valuable when viewed in the light of possible combat in the future. f, as predicted, the cycle of wars continues to emphasize the imited-objective, political-military struggle and to avoid massive dispositions of regular forces, the U.S. Army Special Forces will not have to prove its claim as an exceptionally effective combat unit in he limited conflict.

Starting with a relatively austere organization and lacking clear objectives and co-ordinated support programs, and improvising actics and techniques at widely scattered locations, the U.S. Special Forces in Vietnam achieved notable success. Its members demonstrated repeatedly their combat ability, *esprit de corps*, determination, and willingness to sacrifice. Remarkable is the fact that this evel of effort was just as strong and effective at the end of the ampaign as it was ten years earlier in the beginning.

The conceptual soundness of the organization for Special Forces was tested thoroughly. Though the force was never intended or conducting such programs as the Civilian Irregular Defense Group projects the flexibility of its detachments, the ability of the orce to use support in many forms and from many sources, and he number of specialties represented in each detachment enabled mall detachments operating independently to achieve a variety of bjectives.

Within its own organizational and support limits, the U.S. pecial Forces successfully practiced a number of new tactics and echniques of the highest professional caliber. Notably the force was responsible for the formation and employment of the mobile uerrilla forces—BLACKJACK Operations—and of extended distance econnaissance and security forces—Projects Delta, Sigma, and Omega; the constant circulation of Special Forces resources from acified to contested geographic areas of Vietnam; successful

MEN OF A SPECIAL FORCES A TEAM

operations with other Free World Military Assistance Forces and joint operations with other armed forces; the construction of fighting camps, and in the Mekong Delta region, floating camps; the full-scale employment of irregulars in night operations exclusively; the conduct of waterborne operations in a carefully planned flood campaign, using the Special Forces "navy"—some 400 water craft consisting of airboats, sophisticated U.S. Navy craft, and locally acquired motorized sampans; and the formation and development of airborne-qualified irregular forces as a mobile strike force for use as reserves or as exploitation forces.

The Special Forces founded and operated the theater school, known as the MACV Recondo School, for training reconnaissance troops for all U.S. and Free World forces. It also developed a decentralized form of logistic support featuring direct sea, air, and road shipments to forward supply points in all corps tactical zones. The Special Forces produced a series of handbooks, describing in detail how to carry out any portion of the group's business, from building a camp to serving as an investigating officer. The group developed a civic action program which placed the emphasis on performance rather than philosophy, and on self-help rather than charity; a flexible and controlled accounting system for supplies and funds; and an annual Special Forces campaign plan for utilization of the force in furtherance of announced objectives of the Military Assistance Command, Vietnam. The group also developed a body of lessons learned for review by the Continental Army Command, and, where appropriate, insertion into its training programs.

The Special Forces made recommendations for future doctrine, organization, and equipment. Doctrinally, it was suggested that the mission of Special Forces be expanded beyond the rather narrow speciality of unconventional warfare. Organizationally, recommendations were made to rearrange the detachments to provide greater administrative and logistic capabilities; to add Judge Advocate, Inspector General, Comptroller, and Engineer personnel; to revise drastically the intelligence section; and to give greater responsibility to the noncommissioned officers.

Debits

The U.S. Army Special Forces performance in Vietnam revealed several shortcomings that were constantly under review and analysis, but were still thorny problems at the termination of operations. For instance, psychological operations continued to fail for a number of reasons. There were not enough trained people in the field. Further, the attitude toward integration of psychological operations into tactical plans was indifferent at many levels of command. Direction of the psychological operations effort from the MACV level seemed to emphasize the civic action support theme, to the exclusion of unit level psychological operations tactics and techniques.

Policy direction for the integration of U.S. and Vietnamese psychological operations at the brigade and group level was ambivalent. When guidance did come, it was usually too proscribed to be usable at the lower levels of command. Psychological operations were essentially defensive in nature. Opportunities or suggestions for offensive psychological operations were usually buried in the useless and meaningless statistics of numbers of leaflets delivered or broadcasts made.

The American soldier is the most generous person on earth. It follows that he runs the risk of exhibiting too much concern or extreme paternalism. Since the military and political struggles in Vietnam were being waged simultaneously, the less privileged members of local society made unwarranted assumptions from this display of generosity as to the amount and depth of American support for their cause. The genuine American concern for improving the lot for the underprivileged was given free rein in the early days of the Special Forces in Vietnam; nor was any attempt made by the group to control or limit this generosity firmly up to the time of the group's departure from the country. The sympathy for the minority groups was construed by some as interference to the point that it weakened the American position of rendering advice and assistance to the Vietnamese Special Forces counterparts. Add to this the American characteristic of impatience to get a job done, and the result was a further gap between the Vietnamese Special Forces and the civilian irregulars.

Despite the successful accomplishment of its role of advising, assisting, supporting, arming, clothing, feeding, and shepherding 42,000 irregulars at the peak CIDG strength, and an additional 40,000 Regional Forces and Popular Forces in the local government advisory role, the Special Forces troops were continually conscious of mistrust and suspicion on the part of many relatively senior field grade U.S. military men. This state of affairs, which came about chiefly from a lack of knowledge of Special Forces operations, their limitations and their capabilities, gave rise to many discrete efforts to bring the Special Forces either totally or in separate parts within the operational control of a U.S. senior official. This desire to control Special Forces assets was not restricted to operational commanders, but was evident in many staff officers as well. The most difficult operational control demands came from staff sections at Headquarters, Military Assistance Command, Vietnam. It is also true that some senior Special Forces officers indorsed divided control, since the division of authority, responsibility, and command and control encouraged a situation where both sides, the regular U.S. military chain and the group headquarters, could be played against one another to the advantage of the local U.S. Special Forces commander. To some Special Forces men, the notion of transferring operational control of field detachments to other U.S. Army elements was attractive since it removed a great deal of responsibility for day-to-day operations from the group headquarters. It is a matter of record that the group was most successful when operated as a group, under strong central Special Forces leadership.

Although it has been suggested that had the group been commanded by a general officer many of these travails would have been avoided, there is nothing in the Special Forces experience to validate that speculation. Indeed, if the position of commander were upgraded simply to counter the adverse attitudes, then the more likely consequence would have been continued distrust but under more circumspect approaches. A valid case could be made on an exception basis that the position of group commander exceeded in terms of mission responsibility and liabilities the position of any U.S. brigadier in Vietnam. Should a future commitment of U.S. military forces require the same scale of investment of Special Forces as occurred in Vietnam, the feasibility and desirability of a general officer command of Special Forces should be examined at that time.

The Future of the Special Forces

Certain hard facts have emerged from the experience gained by the Special Forces in the Vietnam War. The Special Forces, for example, can function expertly and efficiently under adverse conditions for long periods of time, as demonstrated by the performances of men stationed in remote camp locations for one-year tours. That Special Forces troops are highly motivated and determined to accomplish their missions as professional soldiers is shown by their repeated tours of duty in the combat zone. The Special Forces organization is very flexible; despite its original focus on unconventional warfare, it adjusted remarkably well to the significantly different methods of countering insurgency by use of conventional forms of warfare with civilian irregulars.

Several questions as to the future role of the Special Forces arose even before the unit departed from Vietnam. Were the tables of organization and equipment adequate and comprehensive? Should the mission of the Army Special Forces be changed to include more than the single mission of waging unconventional warfare? Should there be a permanent branch of service for Special Forces officers and men, as opposed to the detailed, temporary duty nature of their current assignments? Some of these questions are readily answered, some require substantial study.

The tables of organization and equipment have already been modified to incorporate changes brought about by the lessons learned in Vietnam. The new organization has resulted in greater flexibility of employment and more efficient operational capability for the revised group. The doctrinal mission statement of the Special Forces has been revised officially to indicate that the roles and missions are really a function of Special Forces capabilities rather than simply a single unconventional warfare role. This revision permits a broad range of possibilities from the individual in a direct action role to the entire group involved in a guerrilla war.

The question of the feasibility or desirability of authorizing a permanent branch of service for Special Forces officers and men can best be answered by a comprehensive, objective study. The temporary nature of an assignment to Special Forces has created an atmosphere of uncertainty for potential volunteers that has worked to the detriment of the program. Other arms and services quite naturally are reluctant to lose members to the Special Forces for periods up to three years. Within an arm or service, the necessity for formal evaluation of all members of that arm or service within a general pattern of development works to the detriment of the careers of those inclined to Special Forces assignments.

It is impossible to equate combat and command duties between Special Forces officers and their comtemporaries in various arms and services. Special Forces duties and assignments exceed the norm for other arms and services, partly because of the variety of skills and talents embodied in a small force. Quite a few Reserve officers refused to apply for Regular commissions because such action, if approved, took them out of Special Forces immediately and placed them in a career pattern of assignments which were, in their opinion, less appealing than the Special Forces.

As for enlisted men, the education and training necessary to qualify as a Special Forces man led individuals into rather parochial fields. Yet as centralized promotions and proficiency pay criteria became more demanding, the tests for proficiency and standing focused on regular unit performance to the disadvantage of the Special Forces noncommissioned officer. An operations sergeant trained by Special Forces is now competing with an infantry unit operations sergeant through a test mechanism that is focused on the infantry unit, with no allowance for the Special Service unit.

The usual arguments against a permanent branch for Special Forces center on the number of men in the program, the similarity with other combat arms in terms of duties, and a variety of clichés designed to avoid the possibility of setting a precedent for other specialists. If the over-all troop basis of the future contains the permanent feature of Special Forces units, then an analysis in depth should be made to determine the feasibility and desirability of authorizing a permanent branch of service for the members of those units.

One inescapable fact has clearly emerged. The Special Forces men earned on the battlefield their rightful place in the United

States Army. Tough, resourceful, dedicated, and efficient, the men of the Special Forces stood and fought as well and as bravely as those of any fighting unit in our country's history. They are firmly committed to their official motto of "Free the Oppressed" and with equal firmness to their unofficial yardstick; "We are known by what we do, not by what we say we are going to do."

The Special Forces men did their duty well and honorably in Vietnam. They kept faith with the Army and with the United States of America.

Chronology of U.S. Army Special Forces
15 April 1970—1 March 1971

15 April 1970	5th Special Forces Group reduced its personnel space authorization by 400 spaces as directed by U.S. Army, Vietnam.
1 June 1970	Cycle to convert remaining thirty-seven Civilian Irregular Defense Group camps to Army of Republic of Vietnam Ranger camps commenced.
30 June 1970	Detachment B–52, Project Delta, was inactivated. Operation Order 3–70 was published, giving units of the 5th Group instructions for conversion of the Civilian Irregular Defense Group camps to Vietnamese Army Ranger camps.
15 October 1970	5th Special Forces Group reduced its personnel space authorizations by 431 spaces as directed by U.S. Army, Vietnam. The following units were inactivated: the 251st, 487th, 522d, and 540th Signal Detachments and the 585th, 629th, 630th, and 631st Military Intelligence Detachments.
1 November 1970	Company C, 5th Special Forces, completed conversion of camps in I Corps Tactical Zone and retrograded to the Special Forces operations base at Nha Trang for stand-down.
1 December 1970	Company D, 5th Special Forces Group, completed conversion of camps in IV Corps Tactical Zone and was moved to the Special Forces operations base at Nha Trang for stand-down.
7 December 1970	Group S–4 opened a processing point in the logistical supply center to begin disposal of the group's property and equipment.
15 December 1970	5th Special Forces Group reduced its person-

nel space authorizations by 550 spaces as directed by U.S. Army, Vietnam. The 633d Military Intelligence Detachment was inactivated in Vietnam.

19 December 1970 — 5th Special Forces Group was ordered by U.S. Army, Vietnam, to reduce its personnel space authorizations by 710 spaces, effective 15 April 1971. Military Assistance Command, Vietnam, Recondo School, was officially closed.

31 December 1970 — 5th Mobile Strike Force Command, the countrywide Civilian Irregular Defense Group reserve, was disbanded, and its advisory element, Detachment B–55, commenced stand-down. Civilian Irregular Defense Group program countrywide was officially terminated.

1 January 1971 — Company A was placed under operational control of Military Assistance Command, Vietnam, to conduct training missions directed by Commander, U.S. Military Assistance Command, Vietnam.

14 January 1971 — Headquarters, 5th Special Forces Group, was placed under operations control of the deputy commanding general, U.S. Army, Vietnam, and the group was officially notified to stand-down for redeployment not later than 31 March 1971.

15 January 1971 — Company B, 5th Special Forces Group, completed roll-up operations and moved to the Special Forces operations base at Nha Trang for stand-down. 5th Special Forces Group began stand-down for redeployment at the Special Forces operations base.

10 February 1971 — 5th Special Forces Group was notified by U.S. Army, Vietnam, that the plan submitted by the group for redeployment had been approved. An honor guard of 94 men would escort the group colors to the continental United States during the first week of March. One C–141 aircraft would transport the honor guard and selected equipment to Fort Bragg, N.C.

15 February 1971	The following 5th Special Forces Group units were inactivated in Vietnam: the 21st Military History Detachment, 31st Engineers Detachment, and the 588th, 613th, 634th, 641st, and 703d Military Intelligence Detachments.
15 February 1971	All 5th Special Forces Group elements completed turn-in of equipment. 5th Special Forces Group honor guard was selected and began preparations for the redeployment ceremony.
21 February 1971	5th Special Forces Group received official notification from United States Army, Vietnam, that redeployment would occur on 1 March 1971.
24 February 1971	5th Special Forces Group redeployment farewell ceremony was conducted at the Special Forces operation base in Nha Trang.
28 February 1971	5th Special Forces Group operation base was transferred to United States Army, Vietnam.
1 March 1971	5th Special Forces Group honor guard left Nha Trang Air Base for continental United States.

Appendix B

U.S. Army Special Forces Honors

The awards that the men and units of the Special Forces earned in the course of the ten years that the Special Forces served in Vietnam speak for themselves. Twelve men received the Medal of Honor, the nation's highest award for conspicuous gallantry and exceptional heroism under fire:

Ashley, Sergeant First Class Eugene, Jr.*
Bryant, Sergeant First Class William M.*
Dix, Staff Sergeant (later captain) Drew D.
Donlon, Captain (later major) Roger H. C.
Hosking, Master Sergeant Charles E.*
Howard, Sergeant First Class (later first lieutenant) Robert L.
Kendenburg, Specialist Five John J.*
Miller, Staff Sergeant Franklin D.
Sisler, First Lieutenant George K.*
Williams, Captain Charles Q.
Yntema, Sergeant Gorden*
Zabitosky, Sergeant First Class Fred W.

*Awarded posthumously.

In addition men of the Special Forces earned the following number of awards:

Distinguished Service Cross	60
Silver Star	814
Bronze Star	10,160
Purple Heart	2,658
Distinguished Service Medal	1
Legion of Merit Award	235
Distinguished Flying Cross	46
Soldier's Medal	232
Bronze Star with V Device	3,074
Air Medal with V Device	394
Air Medal	4,527

Army Commendation Medal with V Device ..	1,258
Army Commendation Medal	5,650

The 5th Special Forces Group (Airborne), 1st Special Forces, earned the following unit awards in the Vietnam conflict:

Presidential Unit Citation ..	5th Special Forces Group (Airborne), Vietnam, 1 November 1966–31 January 1968
Meritorious Unit Citation ..	5th Special Forces Group (Airborne), Vietnam, 31 January–31 December 1968
Vietnam Cross of Gallantry with Palm	Detachment B–52, Project Delta, 15 May 1964–16 August 1968; Detachment A–322 (Soui Da), 18–25 August 1968; and 5th Special Forces Group (Airborne), 1 October 1964–17 May 1969
Valorous Unit Award	Detachment B–52, Project Delta, 4 March–4 April 1968
Vietnam Civic Action Medal	5th Special Forces Group (Airborne), 1 January 1968–24 September 1970
Navy Unit Commendation Ribbon	Detachment B–52, Project Delta, 17 April–17 June 1967 and 15 July–17 August 1967
Presidential Unit Citation, Navy	Detachment A–101 (Lang Vei), Forward Operations Base 3 (Khe Sanh), and Command and Control (Da Nang), 20 January–1 April 1968

United States Army Special Forces campaign participation credits number twelve for the Vietnam conflict and range from 15 March 1962 to 31 December 1970.

Appendix C

List of Special Forces Camps, 1961–1971

I Corps Tactical Zone (Military Region 1)				
Detachment	Location	Province	Date Opened	Date and Type of Closure
C-1	Da Nang	Quang Nam	Nov 64	Nov 70 converted to ARVN Ranger
Company C	Da Nang (TDY)	Quang Nam	Sep 62	Nov 64 converted to PCS C Detachment
B-11	Chu Lai	Quang Tin	9 Jul 69	Oct 70 closed
B-16	Da Nang	Quang Nam	May 68	1 Jan 70 converted to ARVN Ranger
B-13	Kham Duc (TDY)	Quang Duc	10 Nov 64	26 Jun 65 closed
B-11	Quang Ngai (city)	Quang Ngai	1 Mar 66	10 Oct 66 closed
[a]	A Luoi	Thua Thien	2 May 65	8 Dec 65 closed
[a]	An Diem (TDY)	Quang Nam	Jun 62	11 Jul 64 closed
A-44	A Ro (TDY)	Quang Nam	6 Jun 64	24 Apr 65 closed
A-102	A Shau	Thua Thien	1 Apr 63	12 Mar 66 closed
[a]	Ba Long	Quang Tri	[a]	[a]
A-106	Ba To (new)	Quang Ngai	Mar 65	Sep 70 converted to ARVN Ranger
[a]	Ba To (old)	Quang Ngai	Nov 62	Feb 63 closed
A-110	Con Thien	Quang Tri	Jan 67	25 Jul 67 converted to III MAF
A-100	Da Nang (MGF)	Quang Nam	27 Sep 66	[a]
A-111	Da Nang (MSF)	Quang Nam	Mar 67	1 Jan 70 converted to ARVN Ranger
A-113	Da Nang (TDY) (MSF)	Quang Nam	15 Aug 65	14 Feb 66 closed

A-103	Gia Vuc	Quang Ngai	1 Feb 62	Feb 69 transferred to VNSF
*	Gio Linh	Quang Tri	Oct 66	Dec 66 closed; USMC TAOR
A-104	Ha Thanh (new)	Quang Ngai	28 Apr 65	Aug 70 converted to ARVN Ranger
*	Ha Thanh (old) Son Ha	Quang Ngai	Jan 63	Feb 64 closed
*	Hiep Duc (TDY)	Quang Nam	Apr 63	14 Dec 63 closed
	Hoa Cam (training center)	Quang Nam	Feb 61	10 Oct 66 closed
*	Hue	Thua Thien	1962–63	*
A-105	Kham Duc	Quang Duc	Sep 63	12 May 68 closed
A-101	Khe Sanh	Quang Tri	8 Aug 62	19 Dec 66 converted to III MAF; moved
				to Lang Vei
A-101	Lang Vei	Quang Tri	21 Dec 66	7 Feb 69 closed; moved to Mai Loc
A-101	Mai Loc	Quang Tri	24 Jun 68	27 Aug 70 closed
A-246	Mang Buk	Quang Ngai	Apr 62	Mar 70 converted to Regional Forces
A-108	Minh Long	Quang Ngai	25 Aug 66	Sep 70 converted to ARVN Ranger; 25
		(previous location)	15 Feb 66	25 Aug moved to new location
A-727	Nam Duc (TDY)	Quang Nam	May 62	1 Sep 64 closed
	(Ruong Ruong)			
	(Ta Rau)			
A-105	Nong Son	Quang Nam	24 Jun 68	Oct 70 converted to ARVN Ranger
*	Phu Hoa	Quang Nam	31 Dec 62	Jan 63 closed
*	Phuoc Son (TDY)	Quang Tin	Mar 62	10 Dec 63 closed
FOB	Sa Huynh (MRI MSF)	Quang Ngai	10 Sep 67	27 Apr 68 closed
*	Ta Bat (TDY) (new)	Thua Thien	2 May 65	8 Dec 65 closed
*	Ta Bat (old)	Thua Thien	Mar 63	Mar 64 closed
*	Ta Ko (TDY)	Quang Nam	1 Sep 64	5 Feb 65 closed
*	Ta Ma	Quang Ngai	*	*
A-109	Thuong Duc	Quang Nam	1 Mar 66	Oct 70 converted to ARVN Ranger
A-102	Tien Phuoc	Quang Tin	28 Nov 65	Oct 70 converted to ARVN Ranger
A-107	Tra Bong (new)	Quang Ngai	28 Aug 65	Aug 70 converted to ARVN Ranger; 28
		(previous location)	Mar 65	Aug moved to new location
*	Tra My (TDY) (Hau Duc)	Quang Tin	Oct 62	2 Feb 64 closed
*	Ba Gia	Quang Ngai	*	*

List of Special Forces Camps, 1961–1971—Continued

II Corps Tactical Zone (Military Region 2)

Detachment	Location	Province	Date Opened	Date and Type of Closure
C-2	Pleiku	Pleiku	Nov 64	15 Jan 71 converted to ARVN Ranger
B-22	An Tuc/An Khe	Binh Dinh	Jul 65	1 Jul 66 closed; moved to Qui Nhon
B-23	Ban Me Thuot	Darlac	Nov 64	30 Nov 70 closed
B-24	Kontum	Kontum	Jan 66	31 Jan 70 converted to ARVN Ranger
Company B	Pleiku (TDY)	Pleiku	Oct 62	Nov 64 converted to PCS C Detachment
B-20	Pleiku (MSF)	Pleiku	Nov 67	May 70 closed
B-22 (LN)	Qui Nhon	Binh Dinh	Jan 65	Jul 66 became B-22
B-22	Qui Nhon	Binh Dinh	Jul 66	Jun 69 closed; moved to Chu Lai; became B-11, MR 1
A-334	Ban Me Thuot (TDY) (Eagle Flight)	Darlac	Nov 65	a
A-234	An Lac	Darlac	a	a
A-232	Bao Loc	Lam Dong	30 Sep 65	Jan 66 closed
A-244	Ben Het	Kontum	May 68	31 Dec 70 converted to ARVN Ranger
A-123	Binh Khe (TDY)	Binh Dinh	27 Apr 64	20 Oct 65 closed
A-331	Bong Son (TDY)	Binh Dinh	27 Apr 65	12 Jun 67 closed; moved to Ha Tay
A-424	Buon Beng (TDY) (Cheo Reo)	Phu Bon	May 63	1 Jun 65 closed
A-238	Buon Blech	Phu Bon	Jun 66	Jun 69 converted to ARVN Ranger
A-232	Buon Brieng	Darlac	Mar 64	10 Sep 65 closed
a	Buon Dan Bak	Darlac	Dec 62	Jan 63 closed
A-233	Buon Ea Yang	Darlac	28 Apr 65	3 Aug 66 converted to Regional Forces
A-35	Buon Enao/Buon Yanao	Darlac	4 Dec 61	15 Sep 63 closed
A-233	Buon Mi Ga	Darlac	Dec 62	28 Apr 65 closed
a	Buon Sar Pa	Quang Duc	Jun 63	1 Nov 64 closed

*	Buon Tha (Mewal Plantation)	Darlac	1963	*
*	Buon Ying/Buon Wing	Darlac	Apr 63	10 Aug 64 closed
*	Buon Ya Soup	Darlac	*	Jul 63 closed
*	Buon Yum/Buon Yun	Darlac	Apr 63	24 Apr 64 closed
A-236	Bu Prang (new)	Quang Duc	5 Oct 67	30 Nov 70 converted to ARVN Ranger
A-222	Bu Prang (old)	Quang Duc	Apr 63	15 Jan 65 closed
A-221	Cung Son (new)	Phu Yen	Jan 66	31 Mar 69 converted to Regional Forces
A-313	Cung Son (old) (Son Hoa) (TDY)	Phu Yen	Jun 62	Jan 63 closed; moved to Dong Tre
A-244	Dak Pek	Kontum	Apr 62	30 Nov 70 converted to ARVN Ranger
A-245	Dak Seang	Kontum	Aug 66	30 Nov 70 converted to ARVN Ranger
A-218	Dak Sut	Kontum	26 May 65	18 Aug 65 closed
A-244	Dak To (new)	Kontum	Aug 65	15 May 68 converted to Regional Forces
*	Dak To (old)	Kontum	Feb 63	*
A-411	Dam Pau (Dalat) (TDY) (Serighac Valley)	Tuyen Duc	Aug 62	17 Jan 64 closed
*	Djirai	Lan Dong	Mar 63	*
A-222	Dong Tre	Phu Yen	15 Jun 63	30 Jun 69 converted to ARVN Ranger
A-239	Duc Lap	Quang Duc	Oct 66	31 Dec 70 converted to ARVN Ranger
A-227	Ha Tay	Binh Dinh	12 Jun 67	31 Mar 69 converted to Regional Forces
A-501	Hoai An	Binh Dinh	*	14 Dec 63 converted to MACV
A-221	Kannack	Binh Dinh	Feb 64	2 Jan 66 closed
*	Krong Kno valley	Darlac	Dec 62	Jan 63 closed
*	Krongno	Darlac	*	31 May 63 closed
A-236	Lac Thein	Darlac	20 May 65	30 Sep 67 converted to Regional Forces
A-225A	Le Hai	Phu Yen	*	24 Jun 67 converted to MACV
A-237	Luong Son	Binh Thuan	Jan 66	3 Aug 68 converted to Regional Forces
A-226	Mang Buk (new)	Kontum	Jul 64	31 Mar 70 converted to Regional Forces
A-246	Mang Buk (old)	Quang Ngai	May 62	Aug 63 closed
A-235	Nhon Co	Quang Duc	Mar 66	31 May 70 converted to Regional Forces
*	Phan Rang (Phuoc Thien)	Binh Thuan	May 62	Jul 63 closed

LIST OF SPECIAL FORCES CAMPS, 1961–1971—Continued

Detachment	Location	Province	Date Opened	Date and Type of Closure
A-212	Phi Ho (Le Loi) (Song Mao)	Binh Thuan	8 Nov 65	28 Feb 66 converted to MACV
[a]	Phi Long (Plei Long)	Binh Thuan	[a]	1 Feb 64 closed
[a]	Phi Ma (Plei Me)	Binh Thuan	[a]	5 Dec 63 closed
A-224	Phu Tuc	Phu Bon	21 Apr 65	2 Aug 68 converted to Regional Forces
A-242	Plateau Gi	Kontum	Jan 65	31 May 70 converted to Regional Forces
A-252	Plei Djereng	Pleiku	Jun 64	Dec 66 closed
A-251	Plei Djereng (new)	Pleiku	Dec 66	31 Oct 70 converted to ARVN Ranger
A-254	Plei Do Lim	Pleiku	Apr 62	2 Aug 68 converted to Regional Forces
[a]	Plei Jar	Kontum	[a]	21 Aug 63 closed
A-217	Pleiku (MSF)	Pleiku	Jan 67	[a]
A-218	Pleiku (MSF)	Pleiku	Oct 62	[a]
A-219	Pleiku (MSF)	Pleiku	Oct 62	[a]
A-255	Plei Me	Pleiku	Oct 63	31 Oct 70 converted to ARVN Ranger
A-212	Plei Mrong	Pleiku	Jun 62	1 May 67 converted to VNSF 31 Aug 70 converted to ARVN Ranger
[a]	Plei Ta Nangle	Binh Dinh	Dec 63	8 May 65 closed
[a]	Plei Yit	Pleiku	31 Dec 62	8 Aug 64 closed
A-234	Phey Srunh	Tuyen Duc	Jan 64	1 May 65 converted to Regional Forces
A-241	Polei Kleng	Kontum	8 Mar 66	31 Aug 70 converted to ARVN Ranger
A-112	Polei Krong (TDY)	Kontum	Jan 63	15 Jan 66 closed
[a]	Port Duc Lac	[a]	[a]	[a]
A-223	Qui Nhon	Binh Dinh	Mar 68	Jun 69 closed
[a]	Song Mao	Binh Thuan	May 62	Mar 64 closed
[a]	Suoi Doi	Pleiku	[a]	25 Mar 65 closed
[a]	Tan Canh	Kontum	Aug 62	Feb 63 closed
A-232	Tan Rai	Lam Dong	Jan 66	31 Mar 69 converted to Regional Forces
A-231	Tieu Arar	Darlac	Dec 67	30 Sep 70 converted to ARVN Ranger

Detachment	Location	Province	Date Opened	Date and Type of Closure
A-233	Trang Phuc (Ban Don)	Darlac	Sep 66	30 Sep 70 converted to ARVN Ranger
A-122	Tuy Phuoc (TDY)	Binh Dinh	[a]	25 Nov 65 closed
[a]	Van Canh (old)	Binh Dinh	Nov 62	2 Aug 63 closed
A-223	Van Canh (old)	Binh Dinh	20 Aug 65	14 Apr 66 moved to Van Canh (new)
A-223	Van Canh (new)	Binh Dinh	14 Apr 66	10 Jan 68 converted to Regional Forces
A-228	Vinh Thanh	Binh Dinh	25 Nov 65	25 Dec 67 converted to VNSF 30 Jun 69 converted to Regional Forces
[a]	Ya Lop	[a]	[a]	[a]

III Corps Tactical Zone (Military Region 3)

Detachment	Location	Province	Date Opened	Date and Type of Closure
C-3	Bien Hoa	Bien Hoa	Nov 64	1 Jan 71 closed
B-35	Duc Hoa	Hau Nghia	Dec 66	[a]
B-35	Hiep Hoa	Hau Nghia	Apr 66	Dec 66 closed; moved to Duc Hoa
B-56[b] (Project Sigma)	Ho Ngoc Tau	Gia Dinh	Aug 66	[a]
B-33	Hon Quan	Binh Long	May 65	Nov 70 closed
B-36	Long Hai	Phuoc Tuy	Stp 67	1 Jan 71 closed
B-31	Phuoc Vinh	Binh Duong	[a]	20 Dec 65 closed
B-34	Song Be	Phuoc Long	May 65	13 May 70 converted to Regional Forces
B-32	Tay Ninh	Tay Ninh	Mar 65	Nov 70 closed
B-320	Tay Ninh (TDY)	Tay Ninh	[a]	Mar 65 converted to PCS B-32
A-325	Bao Don	Tay Ninh	1966	Mar 67 closed
A-301	Ben Cat	Binh Duong	Sep 64	31 Aug 70 converted to ARVN Ranger
[a]	Ben Phuoc	Long An	Mar 64	Nov 64 closed; moved to Ben Cat
A-321	Ben Soi	Phuoc Ninh	Mar 65	31 Aug 70 converted to ARVN Ranger
A-302	Bien Hoa	Bien Hoa	Nov 64	[a]
[a]	Bo Mua	Phuoc Thanh	[a]	12 Feb 66 closed
A-341	Bu Dop	Phuoc Long	Nov 63	31 Dec 70 converted to ARVN Ranger
[a]	Bu Ghia	Phuoc Long	[a]	15 Sep 63 closed

LIST OF SPECIAL FORCES CAMPS, 1961–1971—Continued

Detachment	Location	Province	Date Opened	Date and Type of Closure
A-341B	Bu Ghia Map	Phuoc Long	Apr 67	20 Jul 65 closed
A-341	Bunard	Phuoc Long	Jun 63	30 Apr 70 converted to Regional Forces
A-312	Cao Bien	Binh Duong	a	1966 closed
A-333	Chi Linh Cau Song Be	Binh Long	4 Jan 67	15 Dec 69 closed; turned over to 1st Cavalry Division.
A-333	Chon Thanh	Binh Long	a	4 Jan 67 closed
A-342	Dong Xoai	Phuoc Long	May 65	Feb 70 converted to ARVN Ranger
A-351A	Duc Hue (old)	Hau Nghia	a	a
A-325	Duc Hue (new)	Hau Nghia	Nov 67	31 Oct 70 converted to ARVN Ranger
A-343	Duc Phong	Phuoc Long	26 Apr 66	31 May 70 converted ot Regional Forces
a	Go Dau Ha	Tay Ninh	Apr 64	1 Sep 64 closed
A-326	Go Dau Ha (new)	Tay Ninh	1966	a
A-311	Hiep Hoa (old)	Hau Nghia	Feb 63	Nov 63 closed
A-351	Hiep Hoa (new)	Hau Nghia	Apr 66	Dec 67 closed
A-303	Ho Ngoc Tao	Gia Dinh	a	Feb 67 closed; moved to Trang Sup
A-322	Katum	Tay Ninh	Feb 68	31 Oct 70 converted to ARVN Ranger
a	Loc Ninh (old)	Binh Long	Mar 63	a
A-331	Loc Ninh (new)	Binh Long	10 Dec 66	30 Sep 70 converted ARVN Ranger
A-413	Long Thanh (TDY)	Bien Hoa	Apr 63	15 Dec 64 closed
A-601	Long Thanh	Bien Hoa	a	a
A-353	Luong Hoa	Long An	26 May 67	Oct 67 closed; turned over to 51st ARVN Ranger battalion
A-332	Minh Thanh	Binh Long	Dec 63	Mar 70 converted to Regional Forces
A-324	Nui Ba Den	Tay Ninh	Jul 64	Oct 70 closed
a	Nuoc Vang	Phuoc Thanh	Feb 63	1 Apr 64 closed
a	Phuoc Vinh	Binh Duong	a	a
A-322	Prek Klok	Tay Ninh	Mar 67	11 Dec 67 closed
A-322	Suoi Da	Tay Ninh	Apr 65	28 Feb 67 converted to Regional Forces

A–304	Tanh Linh	Binh Tuy	Apr 65	31 Oct 67 converted to Regional Forces
A–323	Trien Ngon	Tay Ninh	8 Dec 69	30 Sep 67 converted to ARVN Ranger
[a]	Thu Duc (training center)	Gia Dinh	Aug 62	[a]
A–334	Tong Le Chon	Tay Ninh	2 May 67	30 Nov 70 converted to ARVN Ranger
A–323	Trai Bi	Tay Ninh	Jun 66	18 Dec 67 closed; moved to Thien Ngon
A–326 (A–352)	Tra Cu	Long An	10 Jan 67	31 Aug 70 converted to ARVN Ranger
A–301	Trang Sup	Tay Ninh	Feb 63	30 Nov 70 converted to ARVN Ranger
A–303	Trang Sup	Tay Ninh	Feb 67	[a]
A–304	Trang Sup (MGF)	Tay Ninh	Jan 67	[a]
[a]	Tuc Trung	Long Khanh	Feb 63	30 Nov 63 closed
A–312	Xom Cat	Long Khanh	Oct 66	26 Mar 67 closed

IV Corps Tactical Zone (Military Region 4)

Detachment	Location	Province	Date Opened	Date and Type of Closure
C–4	Can Tho	Phong Dinh	Oct 64	16 Dec 70 converted to ARVN Ranger
[a]	Can Tho (TDY)	Phong Dinh	Dec 62	Oct 64 changed to PCS "C" Detachment
B–40	Can Tho	Phong Dinh	Mar 65	31 May 70 converted to ARVN Ranger
B–43	Cao Lamh (old)	Kien Phong	Feb 67	7 Apr 69 closed; moved to Chi Lang
B–42	Chau Doc	Chau Doc	1 Mar 65	Jul 68 converted to Regional Forces
B–43	Chi Lang	Chau Doc	7 Apr 69	8 Dec 70 closed
A–432[d]	Chi Lang	Chau Doc	Sep 70	30 Nov 70 converted to ARVN Ranger
[a]	Long Xuyen	[a]	[a]	[a]
B–41	Moc Hoa	Kien Tuong	Feb 65	Oct 70 converted to ARVN Ranger
B–44 (Prov)	Phu Quoc (Xom Duong Dong)	Kien Giang	Feb 67	Jun 68 converted to Regional Forces
A–425[d]	An Long	Chau Doc	1964	26 Jun 66 moved to Thuong Thoi
A–416 (A–221, TDY)	Ap Bac	Kien Tuong	Sep 65	1966 closed
A–424	An Phu	Chau Doc	Mar 64	1 Aug 66 converted to Regional Forces

List of Special Forces Camps, 1961–1971—Continued

Detachment	Location	Province	Date Opened	Date and Type of Closure
A–429 (later A–421)	Ba Chuc (KKK)	Chau Doc	15 Jan 66	Apr 66 moved to Ba Xoai; replaced by new team
A–421	Ba Xoai (KKK)	Chau Doc	May 66	30 Nov 70 converted to ARVN Ranger
A–411	Binh Hung	An Xuyen	Jan 65	9 Nov 67 closed
*	Binh Thang	Kien Tuong	Oct 64	*
A–410	Binh Thanh Thon	Kien Tuong	May 65	31 Oct 70 converted to ARVN Ranger
A–431	Cai Cai	Kien Tuong	Apr 67	30 Sep 70 converted to ARVN Ranger
*	Chau Lang	An Giang	Oct 62	1 Apr 64 closed
A–412	Dan Chu	*	Apr 65	Redesignated A–431 Cai Cai
*	Dan Thanh	Long Xuyen	*	22 Dec 65 closed
A–401	Don Phuc (new)	Kien Phong	Feb 66	31 May 70 converted to Regional Forces
A–430	Don Phuc (old)	Kien Phong	Aug 64	12 Jun 65 closed
*	Du Tho (Doc Trang)	Ba Xuyen	Aug 62	7 Apr 64 closed
A–411	Hai Yen (Song Ha)	An Xuyen	Jun 65	Mar 66 closed
*	Ha Thien (To Chau)	Kien Giang	Feb 63	*
A–421	Ha Tien (new)	Kien Giang	Apr 65	24 Oct 67 converted to Regional Forces
A–412	Kinh Quan II	Xien Tuong	Jan 66	31 Jan 70 converted to ARVN Ranger
*	Long Khanh	Vinh Binh	Feb 63	Jul 64 closed
FOB	Long Khot	*	*	1969 converted to Regional Forces
*	Luong Tam	Chuong Thien	Dec 63	13 Jun 64 closed
A–414	Moc Hoa	Kien Tuong	Mar 63	Mar 68 closed
A–402	Moc Hoa (MSF)	Kien Tuong	1966	Oct 70 converted to ARVN Ranger
A–433	My Da (My An)	Kien Phong	Mar 67	31 Jan 70 converted to Regional Forces
A–416	My Dien II	Xien Tuong	20 Jan 68	30 Mar 70 converted to Regional Forces
A–411	My Phuoc Tay	Dinh Tuong	Feb 67	Mar 70 converted to Regional Forces
A–429	Nui Tuong (proposed site)	Chautoc	15 Jan 66	May 66 abandoned in favor of Ba Xoai
A–441 (A–427)	Phu Quoc (Xom Duong Dong)	Kien Giang	Feb 65	Mar 68 converted to Regional Forces; A Detachment remained until 15 Jun 68

A–442	Phu Quoc (training camp)	Kien Giang	28 Sep 66	15 Jun 68 relocated to To Chau
A–428 (A–311, TDY)	Tan Chau	Chau Doc	4 May 65	1 Jan 67 closed
[a]	Tan Heip	Kien Tuong	Mar 64	[a]
[a]	Tan Phu	An Xuyen	Apr 63	16 Jun 64 closed
[a]	Tan Tu			1964 closed
A–414	Thanh Tri	Kien Phong	Mar 68	31 Aug 70 converted to ARVN Ranger
A–432	Thuong Thoi	Kien Tuong	May 66	Sep 70 flooded, moved to Chi Lang; converted to ARVN Ranger
A–423	Tien Bien	Chau Doc	Apr 64	30 Oct 70 converted to Regional Forces
A–442	To Chau (new)	Kien Giang	Feb 68	31 Aug 70 converted to ARVN Ranger
A–402	To Chau (MGF)	Kien Giang	Feb 67	[a]
A–403	To Chau (MGF)	Kien Giang	Mar 67	31 May 70 closed; property transferred to ARVN Rangers
[a]	Tra Long	Ba Xuyen	[a]	1 Feb 64 closed
A–426	Tri Ton	Chau Doc	[a]	18 Nov 65 closed
A–415	Tuyen Nhon	Kien Tuong	Apr 65	30 Sep 70 converted to ARVN Ranger
A–149 (VNSF) A–432 (USASF)	Vinh Gia	Chau Doc	Aug 64	1 Jul 67 turn over to VNSF; Nov 70 converted to ARVN Ranger
[a]	Vinh Loi	Go Cong	Mar 64	[a]

Special Forces Operations Base

Detachment	Location	Province	Date Opened	Date and Type of Closure
SFOB	Nha Trang	Khanh Hoa	Oct 62 (S–1 & S–4); Feb 63 HQ USASF	Moved from Saigon to Nha Trang
C–5	Ho NGOC Tau	Gia Dinh	1966	[a]
B–51	Dono Ba Thin	Cam Rahn	Apr 64	Converted to MACV
B–50	Nha Trang (Project Omega)	Khanh Hoa	Jul 66	[a]

LIST OF SPECIAL FORCES CAMPS, 1961–1971—Continued

Detachment	Location	Province	Date Opened	Date and Type of Closure
B–52	Nha Trang (Camp Nguyen Van Tan)	Khanh Hoa Khanh Hoa	15 May 64	Aug 70 closed
B–55	Nha Trang (MSF) (originally Nung Security Platoon 1964–65 3d Co MSF)		Nov 64	Dec 70 converted to ARVN Ranger
B–56[b]	Nha Trang/Ho Ngoc Tao (Project Sigma OPCON IIFFV)	Khanh Hoa	Aug 66	[a]
B–57	Nha Trang	Khanh Hoa	[a]	[a]
MACV Recondo	Nha Trang	Khanh Hoa	13 Sep 66	Nov 70 closed
403 SOD	Nha Trang	Khanh Hoa	1966	1970 closed
B–55	Saigon	Gia Dinh	Sep 62	Feb 71 closed
USASF (ABN) Walter Reed Institute of Research Survey Team	[a]	Khanh Hoa	Sep 66	1968 detached from 5th Special Forces Group (ABN)
A–521	Dong Ba Thin	Cam Rahn	1965[a]	[a]
A–503	Nha Trang	Khanh Hoa	Jun 64	[a]
A–502	Suoi Dau	Khanh Hoa	1967	[a]
A–502	Thrung Dung	Khanh Hoa	Mar 64	31 Jan 70 converted to MACV
A–501[c]	Hoai An	Binh Dinh	[a]	14 Dec 64 converted to MACV
A–502	Dien Khanh	Khanh Hoa	Feb 64	Dec 65 closed

[a] Not available or unsubstantiated.
[b] Also listed with Military Region 3.
[c] Also listed with Military Region 2.
[d] Effective 1 June 1967 numbers for Military Region 4 camps were changed to correspond with parent B Detachments as follows:

	Old	*New*
Don Phuc (Mike Force)	A–430	A–401
To Chau (1st MGF)	A–431	A–402
Binh Hung	A–411	A–404
Ha Thien (new)	A–421	A–405
My Phuoc Tay	A–424	A–411
Kinh Quan II	A–416	A–412
Ba Xoai	A–429	A–421
Cai Cai	A–412	A–431
Thuong Thoi	A–425	A–432
My An	A–426	A–433
Phu Quoc	A–427	A–441
Phu Quoc	A–428	A–442

Appendix D

U.S. Army Special Forces Engineer Activities

Engineer support of Special Forces operations from 1961 through 1965 came from three sources: the noncommissioned officer engineer of the A detachment, an engineer construction advisory team, or a naval Seabee technical assistance team.

The engineer noncommissioned officer of each A detachment was basically a demolitionist and usually had no particular training or experience in construction. In some cases, however, he had previously served with conventional combat engineer units where he had acquired essential combat construction skills.

Engineer construction advisory teams provided engineer support on a temporary duty basis from both Okinawa and Fort Bragg, North Carolina. These units were detachments formed under Table of Organization and Equipment 5-500C; each consisted of two officers and two noncommissioned officers equipped with hand tools only. The teams were used primarily in civic action projects such as improving sanitation facilities, schools, and agriculture. That they were obliged to borrow equipment from both the U.S. Operations Mission and the Vietnamese Ministry of Public Works points to an unfortunate deficiency in the table of organization and equipment.

Naval Seabee technical assistance teams, highly skilled, having organic mechanized equipment, and tailored for specific kinds of construction tasks were used during this period with great success. They played a major role in airfield and camp construction.

From 1965 through 1968 engineer support shifted from cellular teams to units of the Free World Military Assistance Forces, as conventional units arrived in Vietnam. Although cellular engineer teams from all sources were used during most of this period, less emphasis was placed upon them because of a reduction in the number of men assigned to temporary duty in Vietnam. By late 1967 the 31st Engineer Detachment, which was organized under Table of Organization and Equipment 5-500E, had arrived in Vietnam, was attached to the 5th Special Forces Group, and was operational. Thereafter it was the primary source of engineer cellular teams for the 5th Special Forces. The engineer detachment was composed of two control teams and four advisory teams and proved deficient for the kind of mission assigned. The mission encompassed all engineering activities in the four corps tactical zones of South Vietnam, including facilities maintenance support for each company as well as backup support in the Special Forces Operational Base. Some of the deficiences were made up by reorganizing the available teams, some by contracting with Eastern Construction Company, Inc., a firm in the Philippines, for certain professional and technical skills. The teams were used as designed and as needed countrywide, with the goal of one in support of each company.

At the peak of engineer activity in the period from the summer of 1968 through the winter of 1971, the engineer of the 5th Special Forces Group was concerned with sixty-nine installations including fifty A detachment camps. The camps often housed more than 1,000 civilian irregulars and their dependents, and had all the attendant problems of communities of such a size. Although very few of these camps were constructed during the period, those that were reflected greater concern for the adequate protection of all the inhabitants. Rehabilitation and improvement projects of the time stressed better protection, water sources in the camps, more adequate dispensaries, and better living conditions in general.

From the latter part of 1969 on, much effort went toward bringing all facilities to a high state of maintenance so that they would require little immediate work when they were turned over to the Vietnamese Army. At the same time the Vietnamese were successfully trained to operate and maintain the installations. The Philippines continued to be the primary source of hard core professional engineering and technical skills absolutely necessary for the construction, operation, and maintenance of relatively sophisticated facilities. Local Vietnamese contractors were used whenever they were available: near the large urban areas it was possible to fill most construction needs efficiently by this means, thereby lessening the demands passed to Free World Military Assistance Forces engineer units.

Engineer functions continued to remain under the staff supervision of the G–4 of the Vietnamese Special Forces high command. Although a Vietnamese Special Forces staff officer was designated to monitor engineer activities during a part of this period, he had no qualified counterparts below the high command level. Most engineer actions were therefore taken unilaterally by U.S. Special Forces after superficial combined planning and co-ordination.

On at least two occasions during this period attempts were

made to convince the Vietnamese Special Forces high command of the necessity for a Civilian Irregular Defense Group engineer unit. The plan called for either a centrally located unit in the vicinity of Nha Trang to be deployed as required countrywide or a unit with each company for use in its own area of responsibility. These attempts were unsuccessful because of the troop ceilings imposed and the unwillingness of the Vietnamese Special Forces high command to trade off some existing spaces for the additional manpower. Since the greatly increased needs generated by the construction of more heavily fortified camps were almost without exception taken care of by engineer units of the Free World Military Assistance Forces, there was no real incentive to make the necessary trade-offs.

When camp construction or rehabilitation projects did make use of CIDG labor, it was on a rotational basis. That is, one third of the men were standing down in camp at any given time—released from their mission—and these men worked on the project. This procedure made it very difficult to execute a training program which would impart some construction skills to the civilian irregulars, but the situation improved considerably toward the end of the period.

The increasing sophistication of facilities at all levels probably peaked with the building of Camp New Bu Prang, occupied by Detachment A–236, on which construction actually commenced in January 1970. The old camp had come under a heavy, persistent, standoff attack ending in late November 1969. The damage sustained led to preliminary planning, including engineering surveys, to determine the scope of reconstruction necessary to make the camp stronger than it had been before the attack. In co-ordination with Headquarters, II Corps, the engineer of the I Field Force, Vietnam, and representatives of the 18th Engineer Brigade, an alternate site was selected in December because there was a likelihood that the II Corps commander would request construction several miles to the southeast of the existing site. Planning was begun for this alternate site upon assurance that all inhabitants of the camp would be fully protected, and a completion date of 15 April 1970 was set. Since the engineer of I Field Force recognized that work of this scope was outside the limited Special Forces capabilities, the project was tentatively assigned to the 18th Engineer Brigade, and thence through the 35th Engineer Group (Construction) to the 19th Engineer Battalion (Combat). The battalion was at this point prepared to start construction at either site.

During the planning process, an understanding that was particularly important to both the Special Forces and the Army Engineers was reached as to the division of work. Basically, the conventional U.S. engineer unit would provide all materials not immediately available through Special Forces channels and would accomplish all equipment work and most tasks requiring special skills on all aspects of construction except for utilities; Special Forces would provide an engineer KB team and, as available, a Detachment A team to advise and assist in the control of a minimum of 100 civilian irregulars per day. These irregulars would do much of the hard labor and unskilled tasks involved in the construction. Special Forces would also drill the wells and furnish the skills and materials necessary to install the electrical, water, sewage, and drainage systems, and the perimeter security wire.

In the latter half of January 1970, the decision was made to construct at the alternate site. Almost simultaneously, necessary orders were issued through both Special Forces and engineer channels. Although planning was far from complete, the initial engineer platoon, from Company D of the 19th Engineer Battalion, was quickly inserted, along with security forces provided from the assets of Company B, Special Forces Group.

At this time several logistical decisions were made. Because of restrictions and a lack of secure roads into the campsite, it was determined that construction materials would be flown into the site. This task was to be accomplished by moving materials from the Cam Ranh Bay Class IV depot to the Cam Ranh Bay Air Force Base and thence to Nhon Co via C–130 aircraft. At that location materials would be prepared for helicopter movement and lifted to the new campsite by Army CH–47 or CH–54 aircraft in a carefully controlled and phased operation.

Another logistical decision was to precutting all bunker materials at the Cam Ranh Bay Class IV depot. They would then be packaged and carefully marked for air shipment in the sequence required. An indication of the scope of the operation is found in the quantity of lumber that was precut and shipped by air to the new campsite—approximately 1.3 million board feet, or 2,000 short tons. This figure is exclusive of shipments of barrier materials, fuel, ammunition, and other supplies and materials.

Construction and final planning proceeded in a parallel fashion into February. During this stage airfield construction commenced along with work on the inner perimeter of the camp. The general plan called for early completion of certain inner perimeter structures, all of which were underground. Work would proceed on the remaining structures within the inner perimeter, the bunkers on the outer perimeter, and the airfield.

Progress remained on schedule into February because a relatively low level of productivity was assumed during the early stages of work. However, productivity did not increase as scheduled in the latter part of February and onward due to equipment problems and the drain on air assets during the Cambodian incursion which commenced in March. Also, design changes on the airfield and construction deficiencies in the camp itself caused an extension of the schedule.

The 15 April completion date was established for two reasons: the first dealt with tactical considerations in that available forces were spread very thin protecting two campsites. The second very practical reason was that the monsoon rains could be expected to commence around that time. When the rains arrived, there was still much to be done that was directly hampered by the ensuing muddy conditions. Finally, after very slow progress, an unseasonal break in the weather provided the necessary respite, and the work was complete in early August 1970.

This camp was the epitome of a fighting camp, well sited on a commanding hill with all facilities underground. The fighting bunkers were placed so as to maximize the effect of defensive fire and at the same time protect civilian irregulars and their dependents against either direct or indirect fire. Because of its sophistication and isolation, construction of this camp proved to be a real challenge to the capability and ingenuity of all concerned.

Appendix E

After Action Report MIKE Force: Attleboro, 1–7 November 1966

SUBJECT: After Action Report MIKE Force/"Attleboro" 1-7 Nov 66

General: a. Third Corps MIKE Force had moved to Loc Ninh on 30 October 1966 in support of moving to new camp site, and was on an operation in Loc Ninh area.

b. Third Corps MIKE Force was alerted 2 November 1966 to move from Loc Ninh to Suoi Da. The move was completed at 1430 2 November 1966.

3. Task Organization
   a. 530 Nungs in three (3) companies.
   b. Seven USASF EM.
   c. One USASF Officer.
4. Mission: Combat Reconnaissance.
5. Sequence of Events:

1 Nov 2210—China Boy alerted for movement from Loc Ninh to Suoi Da.

2 Nov 0800—1st MIKE force company extracted from LZ.

0900—Company closed Loc Ninh.

1045—2nd Company extraction began.

1215—Completed extraction 2nd Company.

1030—C-123 aircraft began arriving Loc Ninh.

Direct support helicopter company moved from Loc Ninh to Tay Ninh East to lift MF from Tay Ninh East; and to lift MF from Tay Ninh West to Suoi Da.

1352—Tay Ninh.

1530—Movement of MF from Tay Ninh to Suoi Da completed.

2 Nov 1630—China Boy Company 3 deployed.

3 Nov 0830—China Boy Company 1 deployed.

1220—China Boy Company 3 engaged est VC Co vic XT486687. VC broke contact 1245, fled north. SSG Monaghan wounded right arm and fingers (GSW).

1815—China Boy Company 1 made contact vic XT458587 with est VC platoon. VC broke contact 1830. SSG Garza WIA (GSW).

4 Nov 0730—China Boy Company 1 hit mined area vic XT561588. One MF KIA, two MF WIA. Medevac chopper downed by SF fire vic XT485622, while enroute to China Boy Company 1's location. One US KIA (Crew Member), chopper was recovered.

1445—China Boy Company 3 made contact vic XT416670 with est VC Bn or Regt. China Boy Company 3 withdrew south and called in airstrike. On initial contact chain saws, generators, and trucks could be heard. VC counterfired with 81mm and 60mm mortar, AW and SA fire, then tried to close with China Boy Company 3 elements.

1800—Received resupply of ammunition and food vic XT435668.

2200—China Boy Company 3 indicated that he was receiving heavy casualties and VC were encircling him.

2300—China Boy Company 3 indicated light contact.

5 Nov 0230—China Boy Company 3 indicates contact with VC has ceased.

0730—China Boy Company 3 receiving heavy volume of fire. Requested reinforcements.

0745—Radio contact with China Boy Company 3 broken.

0800—China Boy Companies 1 and 2 proceeding to China Boy Company 3's location.

0845—China Boy Company 2 hit VC bunkers. Negative contact.

0940—28 MF personnel picked up by CIDG CO from Suoi Da.

0935—China Boy Company 3 having casualties evacuated vic XT388634.

1200—Three USSF MIA.

55 MF from China Boy Company 3 made linkup with China Boy Companies 1 and 2. Of those 15 to 25 WIA.

5 Nov 1330—One MF drowned while crossing river with China Boy Company 1.

6 Nov 1040—Est VC platoon with mortars attacked Suoi Da airfield, 4 CIDF KIA, 2 WIA.

1700—9 MF personnel closed in to Suoi Da.

7 Nov 1430—MF begins move to Loc Ninh.

1600—MF completes move to Loc Ninh.

1645—One US body found.

1830—MF bodies returned to Bien Hoa by CV-2 aircraft.

6. (C) On 30 October, all 3 companies of the Mike Force deployed to Loc Ninh, A-331 Binh Long Province, to conduct operations in response to intelligence reports that the camp was a possible target for a major VC attack prior to 11 November 1966. However, hard intelligence reports received on 1 November indicated that a VC regiment had moved into the operational area of Camp Suoi Da, A-322, Tay Ninh Province. A decision was made to move the Mike Force into that area, and this was accomplished on 2 November. On 031220 November, the 3rd Mike Force Company made contact with an estimated VC company. The VC immediately broke contact and an airstrike was called in on their route of withdrawal. At 031845, contact was again established with an estimated VC platoon which resulted in 10 VC KIA and two USASF WIA. At 040730, the 1st Mike Force Company hit a mined area and suffered one KIA and two WIA. A Med Evac chopper in the same general vicinity was shot down by small arms fire and resulted in one US KIA. At 041445, the 3rd Mike Force Company made contact with an estimated battalion or regimental sized VC force. This contact resulted in 15 VC KIA and two Mike Force WIA. The Mike Force Company was still in contact at 041540 and attempted to withdraw to the south. At 042000 the Mike Force Commander reported that he was surrounded and had suffered 35 casualties (KIA). The remaining two CIDG companies departed Camp Suoi Da to reinforce the operation. At 042305 the 3rd Mike Force Company Commander reported that he was still in contact. Enemy casualties reported at this time were 50 VC KIA. Contact with the VC was broken at 050330. At 050900 the 3rd Mike Force Company again reported that they were receiving a heavy volume of fire. They were instructed to secure an LZ so that an attempt could be made to extract them from the area. This was accomplished at 051200. All Mike Force elements were extracted at 051830.

Interview with SFC Heaps, 7 Nov 66

At 021630 Nov China Boy 3 landed at LZ vicinity XT491644, and began moving north (see attached overlay). At 031220 Nov vic XT473683 China Boy 3 discovered tunnel complex and fortifications. While destroying complex, VC fired on China Boy 3 wounding SSG Monaghan. China Boy 3 withdrew east to LZ, vic XT487686. Med Evac arrived, casualties were loaded, but because chopper was overloaded it could not take off. SSG Hunt, who came in with the Med Evac elected to remain with China Boy 3 so that casualties could be evacuated. Again China Boy 3 moved west to tunnel complex, but could not take it because of intense fire. China Boy 3 broke contact and moved to vicinity XT465692 (see overlay). Here they heard several motors that sounded like generators and trucks, plus several chain saws. Heavy contact was made. China Boy 3 received heavy automatic weapons fire and mortar fire. Mortars sounded like 60mm. China Boy 3 broke contact and moved to LZ vicinity XT435667 (see overlay). All during the time they were moving to LZ they were receiving sporadic small arms fire. Also when they crossed road vic XT453667, they received mortar fire.

At LZ vic XT435667 China Boy 3 received resupply of food and ammunition. From resupply LZ the unit moved to vicnity XT444672, went into defensive perimeter and began breaking down ammo. While they were breaking down ammo, the VC attacked from the east in a "U" shaped formation. It was beginning to get dark and the VC withdrew to approximately 100 meters east of China Boy 3's position and maintained contact all night. At approximately 0645-0700 the following morning the VC made another assault on China Boy 3's position and overran them. SFC Heaps and SSG Hunt were wounded during this assault. SFC Heaps said he was knocked unconscious and when he came to SSG Hunt was giving him first aid. Heaps and Hunt decided to get to the LZ vic XT424680. They had two Mike Force with them, one was wounded. They couldn't move very fast or very far without resting, and Heaps and Hunt would pass out periodically. Finally Hunt said he could go no further so Heaps left one Mike Force with Hunt and continued to the LZ. After this Heaps didn't remember anything.

Interview of SFC E7 Heaps on 7 Nov 66, 3rd Field Hospital.

REFERENCE: LOC NINH Map Sheet 6245 II
465691 to 473681, first contact

Area of Operations: Major contact overrun at 440669

Weather: Excellent, high clouds, temperature

Terrain: Jungle, (thick) close to water supply, within 500 meters of road on high ground.

Fortifications: Tunnel and bunker complexes for one, two or squad size positions. All with overhead cover and pre-arranged fields of fire. Positions were hardened against direct fire.

Weapons, uniforms and equipment: Automatic weapons were in abundance; of the two weapons captured they were AK's. They had a lot of machine guns, sounded like 30 cal, heavy. Uniforms were mixed, personnel KIA had on black shoes. All of the soldiers encountered had complete sets of web gear.

Significant weapons: Grenade launcher which looked like our "LAW," light in weight, approximately 3 feet long, markings appeared to be Chinese, possible identification: Chinese antitank grenade launcher type 56, P. 155 DA pamphlet 381-10. Indirect fire was provided by 60mm mortars, identification by rounds.

Tactics: Fire discipline was excellent. Upon making contact, VC fired in mass; upon breaking contact VC ceased fire without sporadic firing. The VC maintained contact while the unit was trying to break contact.

They mortared and sniped at them in the retreat. After fixing the new location of the 3rd Company they (VC) attacked using squad fire and maneuver up to grenade range and then reverted to individual action. By this time, it was almost dark, so the VC withdrew approximately 100 meters and maintained contact all night. At 0645-0700 the next morning they assaulted using the same tactics with a heavy volume of fire suppressing the 3rd Company's position.

Movement was forward by flanks and frontal assault forces.

Other: The VC troopers were young and aggressive.

At grid 465691—Generators and chain saws were heard.

US/Mike Force and VC casualties from 2-11 Nov, Opn "ATTLEBORO"

UNIT	FRIENDLY			UNIT	VC		
	K	W	M		K	POSS	POW
1st	22	72		1st	721		5
196th, 25th	46	300	5	25th	181		
USASF	1	3	1	173rd	4		
Mike Force	25	17	7	Mike Force	85	448	
TOTAL:	94	392	13		991	448	5

Following units identified:

271, 272, 273, 320 MFPMB, 70 Regt, 10th NVA Inf.

271 by contact and/or documents.

273 Regt and 272 Regt contact XT-4254 by 2/1st Inf.

272 Regt contact on 4 Nov at Suoi Cau XT-4530 one of our agents says the 320 are also involved. This is reasonable since this whole area is in their AO.

101st Regt by captive XT431559 on 8 Nov.

70 Regt possibly contact by 1/16 Inf on 9 Nov.

271 by 1 PW on 3 Nov, this unit identified by ASPAR in the southern part of the contact area SW of Dau Tieng.

All of the documents and PW intell reports are still at lower level Headquarters and have not reached FFII yet for closer examination.

US units committed to Operations.

US 173d Abn Bde—2 Bns

ARVN Ranger Bns (Attached)—2

US 1st Div—8 Bns

US 25th Div—3 Bns

196th Bde—1 Bn

/a/ Thomas Myerchin
/t/ Thomas Myerchin
CPT, INFANTRY
COMMANDING

Appendix F

After Action Report: Operation Blackjack 33

1. SIZE AND COMPOSITION OF OPERATION: Reconnaissance Company with company headquarters; Nine Roadrunner Teams composed of four Vietnamese Nations (VNN) each; Seven Reconnaissance teams composed of two USASF and four VNN each; Three Mike Force Companies composed of 11 USASF and 428 VNN; Detachment A-303/Mobile Guerrilla Force 957 composed of 13 USASF and one Mobile Guerrilla Company, 174 VNN.

2. MISSION: To conduct extended reconnaissance and Mobile Guerrilla Force operations in AO Blackjack as directed by CG, 1st Infantry Division.

3. TIME OF DEPARTURE/RETURN: Operation Blackjack 33 was conducted from a Forward Operations Base (FOB) located at Phuoc Vinh, RVN inside the perimeter of 1st Bde, 1st Inf Div, during the period 24 Apr-24 May 67. Project SIGMA advance party began movement by C-130 from Bien Hoa Airbase to FOB 240800 Apr and completed movement 241700 Apr 67. FOB activated 241730 Apr. The main body was moved by C-123 transport from Bien Hoa Airbase with the first element arriving 261000 Apr and the last element closing at 261630 Apr 67. The operation terminated 240700 May and the Detachment command element (-) was returned to Base Camp by helicopter closing 240800 May. The main body began movement 240820 May by C-130 transport to Bien Hoa Airbase. Project SIGMA elements closed at Base Camp 241548 May 67.

4. CONDUCT OF OPERATIONS: Operation Blackjack 33 began with the issuance of a verbal OPORD on 191300 Apr 67. CO, Det B-56 issued a fragmentary order for movement of an advance party to Phuoc Vinh, RVN, on 24 Apr 67. On 20 Apr 67 the units airlift requirements were submitted by the S3 to II FFV. An advance coordination party departed for Phuoc Vinh to select a unit location. The S2 visited II FFV, CICV, and G2, 1st Inf. Div and obtained current intelligence and requirements concerning the assigned AO. A message was received placing Project SIGMA in direct support of the 1st Inf Div effective 25 Apr 67 for period of 60 days. On 21 Apr, CO, Det B-56, CO, Recon Company, S2, Det B-56 and CO, MGF 957 made a coordination visit to Hq, 1st Bde, 1st Inf Div. Det S3 published a unit movement order for

Operation Blackjack 33. OPORD 8-67 (Operation Blackjack 33) was published 242300 Apr 67. On 271400 Apr an aerial reconnaissance of the entire AO was conducted. Helicopter support, provided by 162d Assault Helicopter Company, was composed of four transport and two armed helicopters. Two O1E radio relay were provided by 184th Avn CO. CO, Det B-56 ordered three Roadrunner Teams to begin preparation for infiltration on 27 Apr. On 281055 Apr, CO MGF 957 arrived at the FOB and briefed the CO, Det 3-56. On 281345 Apr, CO, Det B-56 and CO, MGF 957 briefed the CO, 1st Inf Div and on 290800, S3, Det B-56 briefed CO, 1st Bde and staff. On 021907 May MGF 957 (-) was lifted by helicopter into AO vicinity Chi Linh. On 022020, CO Det B-56, DCO and S3 returned from visit with MGF 957 at Trang Sup and Chi Linh. MGF 957 was extracted from AO with KIA and MIA and closed FOB 032030 May. On 040830 May, CO, 5th SFGA and DCO, Spec Opns, arrived FOB for briefing on MGF 957 engagement. On 041145 May MGF 957 was moved to Trang Sup by 2 CH-47 helicopters. On 071500, CO, MGF 957 presented mission briefback at FOB for DCO Spec Opns, 5th SFGA and party. MGF 957 closed at Dong Xoai 091700 May and entered AO on foot at 092400 May 67.

a. Reconnaissance Team Operations:

(1) Recon Msn 1: Team received order 290930 Apr and presented mission briefback 29200 Apr. Team infiltrated 300626 Apr vic XT971400 without incident. At 301025 Apr team sighted approx 5-10 VC vic XT971679 and requested artillery at 301049 Apr. Results of artillery fire are unknown. Team was extracted by sling under fire vic XT988682 at 011810 May.

(2) Recon Msn 2: Team received mission order 291000 Apr and presented mission briefback 292030 Apr. Team attempted infiltration vic XT975397 at 300759 Apr. Team leader suffered a broken leg exiting the helicopter and the mission was aborted.

(3) Recon Msn 3: Team received mission order 302000 Apr, conducted aerial recon of RZ 010900 May and presented mission briefback 012030 May. While attempting infiltration vic XT918571 at 020620 May, the helicopter crashed, not as a result of enemy action. Four USASF personnel were slightly injured, but all returned to duty. Gunner of helicopter crew was medevaced. All personnel, radio and weapons were immediately evacuated. A reaction platoon was airlanded at 020710 May and secured the downed helicopter until evacuated.

(4) Recon Msn 4: Team received mission order 011400 May, conducted aerial recon of RZ 020900 May and presented mission briefback at 021300 May. Team was infiltrated vic YT012814 at 030650 May. At 030715 May 2 VC were sighted and a hasty ambush was set up in an attempt to capture a PW. The two VC sighted returned in approx 20

minutes with estimated 25 men and team heard signals from three sides. Team requested exfiltration and moved to a LZ vic YT008813 and were extracted by sling at 030850 May under fire.

(5) Recon Msn #5: Team received mission order 042000 May, conducted aerial recon at 051400 May and presented mission briefback 051900 May. Team infiltrated vic XT994563 at 061159 May without incident, and exfiltrated vic XT980568 at 071205 May when two VNN team members became too ill to continue. No significant sightings.

(6) Recon Msn 6: Team received mission order at 052000 May, conducted aerial recon of RZ at 061000 May and presented mission briefback 061900 May. Team infiltrated vic YT022582 at 070952 May without incident, and was exfiltrated vic YT010613 at 080740 May due to faulty radios. Team was re-infiltrated in vic YT036633 at 081339 May without incident after securing a new radio at the FOB. Team moved to a well used trail vic YT036638, turned west on trail for 100 meters then observed and killed one VC (BC). Team called for exfiltration and was extracted by sling vic YT039639 at 081445 May.

(7) Recon Msn 7: Team received mission order 062000 May, conducted aerial recon of RZ 071000 May and presented mission briefback at 071600 May. Team infiltrated vic YT061372 at 080730 May and moved crosscountry until 081550 May at which time they observed a VC jump from a tree vic YT051700. Team assumed position was compromised and requested exfiltration which was completed from vic YT052200 at 081605 May without incident.

(8) Recon Msn 8: Team received mission order 071000 May, conducted aerial recon of RZ 071500 May and presented mission briefback at 082030 May. Team infiltrated vic YT030440 at 090655 May. Team observed an estimated 200 VC moving southwest. Airstrikes were called in and a ground assessment by two platoons of 1st Company reported 43 VC KBA (BC), 9 VC KIA (BC) and 2 VC CIA and a large amount of medical supplies recovered. Two team members became separated due to a misunderstanding and were exfiltrated vic YT055414 at 111130 May. Remainder of team was exfiltrated from LZ at 111145 May.

(9) Recon Msn 9: Team received mission order 082000 May, conducted aerial recon of RZ at 091300 and presented mission briefback at 091700 May. Team infiltrated vic XT 946404 at 100705, moved only 100 meters and found themselves surrounded by an estimated two platoons of VC. At 100750 May team was receiving fire from three directions. Gunships suppressed the area with fire and the team was extracted at 100815 May. Artillery fire was placed in area with unknown results.

(10) Recon Msn 10: Team received mission order 091600 May,

conducted aerial recon of RZ 100800 May and presented mission brief-back at 101230 May. Team infiltrated vic YT023711 at 101558 May without incident. Team moved immediately into an ambush position with the mission of capturing a POW. Team remained in ambush position with negative contact and was exfiltrated from vic YT023705 at 111732 May. Team was exfiltrated because one VNM team member was believed to have appendecitis.

(11) Recon Msn 11: Team received mission order 110800 May and conducted aerial recon of RZ 111500 May and presented mission briefback at 120800 May. Team infiltrated vic YT036634 at 121340 May with the mission of capturing a PW. Team set up ambush on a trail vic YT039639 and at 121510 May ambushed one VC, wounding him. Team then called for extraction but POW DOW before they were extracted from infiltration LZ at 121530 May without incident.

(12) Recon Msn 12: Team received mission order 131000 May and conducted aerial recon of RZ 131500 May and presented mission briefback 141700 May. Team had planned infiltration for RZ HOTEL, but team leader was medevaced with a perforated eardrum. A new team leader was assigned and team infiltrated RZ GOLF vic XT945348 on 141458 May. Team captured one VC (female) nurse in close proximity to infiltration LZ and was exfiltrated from that location at 141528 without incident. PW was turned over to 1st Inf Div for interrogation.

(13) Recon Msn 13: Team received mission order 141000 May, conducted aerial recon of RZ 141500 May and presented mission brief-back 142000 May. Team infiltrated vic XT981651 at 151245 May. Team encountered a VC force of 7 men at 151500 May and opened fire, killing 2 VC (BC). Team, was extracted under fire at 151545 May from infiltration LZ.

(14) Recon Msn 14: Team received mission order 151000 May, conducted aerial recon of RZ 161430 May and presented a mission briefback at 161910 May. Team infiltrated vic YT045645 at 170655 May and moved to a well used trail to observe and capture a PW. Team observed an estimated VC Battalion using the trail, artillery fired with unknown results and team was extracted vic YT036644 at 181510 May without incident.

b. Roadrunner Team Operations.

(1) RR Msn 1: Team was infiltrated vic XT687572 at 271617 Apr without incident. Team made contact with 3 VC vic XT684568 at 271630 Apr. Team was fired on vic XT678527 during movement to LZ. One man WIA, left arm (DOW) and team extracted under fire at 281710 Apr vic XT665509.

(2) RR Msn 2: Team was infiltrated vic XT614525 at 271655 Apr without incident. Observed 4 VC at XT625546 with machettes.

Team fired on from church vic XT658579. Team threw grenade into the church with unknown results. Team suffered negative casualties, and was exfiltrated 281707 Apr vic XT084590 without incident.

(3) RR Msn 3: Team was infiltrated via XT615630 at 271640 Apr. Team moved approx 200 meters off the LZ and were observed by 6 VC. Fire was exchanged with unknown results. Team threw emergency smoke and were extracted under fire at 271700 Apr vic of the infiltration LZ. Negative friendly casualties.

(4) RR Msn 4: Team was infiltrated 281902 Apr vic XT688444 without incident. Team observed a VC platoon (40 men) at 282130 Apr vic XT833521 moving south on road. Team contacted 8 VC at 290910 Apr vic XT844479 and were followed and fired on with negative casualties. Team was extracted at 291647 Apr vic XT837473.

(5) RR Msn 5: Team was infiltrated at 281855 Apr vic XT840525. Team sighted one VC Platoon moving south on road vic XT641489 at 291400 Apr. VC fired at team at 291415 Apr and wounded team leader and one other team member. Team returned fire and killed 4 VC. Team evaded and were extracted at 291657 Apr vic XT632487

(6) RR Msn 6: Team infiltrated 291808 Apr vic XT944633 without incident. Team made contact with approx one VC Platoon at 300745 Apr vic XT896748. VC were in foxholes and fired at the team, team didn't return fire and were exfiltrated at 301705 Apr vic XT917735 without incident.

(7) RR Msn 7: Team was infiltrated at 291759 Apr vic XT944633 without incident. One man became lost from team upon infiltration and other three members became lost during movement. Team of three was fired on vic XT949631. Team returned fire with unknown results. Team encountered one VC at 0292300 Apr vic XT954630 in a hut and estimated a VC Platoon later tried to locate them with flashlights. Entire team joined up and were extracted by sling at 300705 Apr vic XT957768 without incident.

(8) RR Msn 9: Team infiltrated at 301840 Apr vic XT889608 without incident. No contact or significant observations were made and team was exfiltrated at 011709 May vic XT916511 without incident.

(9) Msn 10: Team infiltrated at 301845 Apr vic XT924345 without incident. Team contacted an unknown number of VC vic XT933356, who fired approx 15 rounds with an automatic weapon. After team retreated, VC fired single shots and signaled during the night, firing two shots every hour. Team was exfiltrated at 011658 May vic XT931356 without incident.

(10) RR Msn 8: The scheduled infiltration of RR Msn 8 at 301830 Apr was aborted due to weather and was later infiltrated at

010739 May vic XT978625 without incident. Team had negative contact and was exfiltrated at 011719 May vic XT939564.

(11) RR Msn 11: Team infiltrated at 011734 May vic YT101824, moved to a trail where 3 VC were observed at 011740 May vic YT073825. VC signaled with two shots, the team answered with one shot and the VC moved away. AT 011830 May vic YT056815 the team observed a Vietnamese dispensary with 3 VC guards, who fired on the team. Team withdrew to a CIDG OP vic YT078765 at 011950 May, then moved to Dong Xoai. Team was exfiltrated at 021040 May vic YT088757 without incident.

(12) RR Msn 12: Team infiltrated at 011739 May vic YT056854 without incident. 4 VC fired on team vic YT073912. Team returned fire with unknown results. Team moved to the USASF Camp Dong Xoai and were exfiltrated at 021600 May.

(13) RR Msn 13: Team infiltrated 021600 May vic YT1081759 without incident. Team sighted 5 VC YT149843 without being detected. Team was exfiltrated at 031915 May vic YT136068.

(14) RR Msn 14: Team infiltrated at 050802 May vic YT023554 without incident after infiltration of 03 May was cancelled due to increased activity of MGF 957 and aborted on 04 May due to weather. Team made no contact or significant sightings. Team was extracted by vehicle at 051150 May vic XT995522, after contacting a friendly PF post.

(15) RR Msn 15: Team infiltrated at 050808 May vic YT074417 after being delayed for the same reasons as RR Msn 14. Team returned to FOB by truck 051400 May.

(16) RR Msn 16: Team infiltrated at 051403 May vic XT930535. Helicopter received approx 10 rounds all fire departing the LZ after infiltration. Team was exfiltrated 061008 May vic XT914573 without incident.

(17) RR Msn 17: Team was infiltrated 051402 May vic XT960530 and exfiltrated 061012 May vic XT983573 without incident.

(18) RR Msn 18: Team infiltrated 061202 May vic YT027529 and VC immediately fired three warning shots and began signaling one another. Team retreated south for 300 meters, then captured a VC suspect vic YT013533 at 061330 May. Team with suspect were extracted vic XT993519 at 061415 May.

(19) RR Msn 19: Team infiltrated vic XT956594 at 071414 May. Immediately on leaving the helicopter and moving north into the tree line they observed 20 VC to their front in foxholes with overhead cover. The VC opened fire and told them to throw down their weapons. Team returned fire killing two VC (BC) and one team member slightly

WIA. Gunships suppressed the VC and team was exfiltrated from the same LZ at 071426 May.

(20) RR Msn 20: Team infiltrated vic YT062719 at 101136 May and was extracted vic YT028629 at 101605 May. Team encountered 9 VC vic YT033638 at 101555 May.

(21) RR Msn 21: Team infiltrated vic YT062342 at 120851 May. At 121200 vic YT019346, team sighted 4 VC who fired two rounds, but team did not return fire. At 130830 May team observed 2 VC vic YT012282. Team was exfiltrated vic YT011284 at 13100 May.

(22) RR Msn 22: Team infiltrated vic XT938346 at 140807 May and was extracted by rope ladder vic XT973389 at 151000 May.

(23) RR Msn 23: Team infiltrated vic XT941622 at 150837 May. Team observed 30 man VC Platoon vic XT956637. At 151430 May, team was observed by estimated 2 VC Platoons vic XT963363. VC fired a rifle grenade near the team then began advancing without firing. One team member threw a M-26 grenade when the VC were about 15 meters away, and killed 5 VC (BC). Team then took cover in two old foxholes and killed 4 additional VC (BC). Gunships suppressed fire and team was extracted under fire vic XT963636 at 151525 May.

(24) RR Msn 24: Team was infiltrated vic YT015262 at 160934 May. Team fired upon by estimated 14 VC at 161000 May vic YT024264. Team exfiltrated vic YT034266 at 161512 May.

(25) RR Msn 25: (8 man team): Team infiltrated vic XT972623 at 190903 May. Team made contact with 3 VC vic XT966651 killing one VC (BC) and capturing one homemade weapon. Team then moved towards the LZ and were fired upon by VC with 3 rifle grenades. Team made contact with interpreter in the radio relay aircraft at 191155 May and requested immediate extraction from vic XT962638. During the recovery of this team two helicopters were downed by enemy fire. Two USASF WIA, 1 US pilot WIA, 1 interpreter WIA and 1 Roadrunner WIA. Downed aircraft were secured by 1st Company and MGF 957 and later recovered.

(26) RR Msn 26: Team infiltrated vic XT845449 at 210820 May and moved a short distance when they discovered 7 VC following. Team fired, killing 2 VC (BC) and requested extraction. Team extracted vic XT848451 at 210840 May with negative friendly casualties.

(27) RR Msn 27: Team infiltrated vic XT823498 at 211312 May. Team made contact with 4 VC vic XT829497 at 211420 May. They exchanged fire and VC disappeared. Team was extracted under fire at point of enemy contact at 211428 May. Team suffered two cases of heat exhaustion.

(28) RR Msn 28: Teams infiltration was cancelled enroute to

LZ due to supporting Aviation unit being returned to control of 1st Bde, 1st Inf Div for another mission.

c. Mike Force Company Operations:

(1) 2d Mike Force Company: The company was alerted on 1 May to move to Dong Xoai with the mission of acting as a reaction force at that location. Unit closed Dong Xoai 020847 May. On 050500 the unit departed Dong Xoai on a reconnaissance in force mission. The patrol apprehended 2 VCs vic YT074834. At 070800 May an estimated VC platoon was observed vic YT073843 moving east. Company fired on VC and they fled east without returning fire. The Company received AW fire 070805 May from vic YT074826 by unknown number VC. Company had 4 WIA and requested a medevac. Company covered approx 40 KM on patrol and closed at Dong Xoai 071445 May. VCs were turned over to LLDB for interrogation. Company was moved from Dong Xoai to FOB by CV-2 aircraft and closed FOB 072145 May.

(2) 3d Mike Force Company: Received mission on 061400 May from S3, Det B-56, gave briefback for 3 day mission in RZ CHARLIE at 071900 May. At 080800 May company departed from vic YT045490 on foot. Unit moved east for 7000 meters and set up a patrol base vic YT030490 and conducted local patrols with negative results. At 081800 May 5 shots were fixed about 300 meters west of the base along the route company had moved in on. A patrol was sent to the vicinity of the shots with negative contact however. On 090700 May company moved from patrol base and at 090800 May 1 shot was fired from area of the patrol base. Again a patrol was sent out to the area, but with negative contact. Company moved on for 500 meters and found 2 graves vic YT038470. Graves contained 2 bodies in body bags, approx 2 months old. Unit moved south at 091150 May and vic YT041450 set up second patrol base. Ambushes were set along Song Be River with negative contact. Company moved from night location 100700 May. At 100900 May another body was found vic YT034456, similar to first two. Company closed FOB at 101500 May.

(3) 1st Mike Force Company: Company was infiltrated into RZ ECHO by helicopter at 121100 May vic XT876578. After company plus recon element landed and was formed, recon element, took point and moved from LZ on 120 deg AZ. Unit came upon main trail vic XT887565. Unit moved for approx 600 meters and stopped for noon meal. At this time fire was heard from vic XT8755 at 1430 hrs. Fire was directed at FAC aircraft. At 1700 hrs unit stopped for night at XT887538. Unit moved out 130645 May continuing west along trail. At trail junction vic XT870537, unit stopped while numerous trail complexes were reconned. Upon rejoining unit, recon element took point and unit moved SE. At 131730 unit stopped for night vic

XT903487. Unit moved south at 140700 May and was resupplied at XT925417 at 141330 May. After re-supply unit continued south. At approx 1600 hours unit began moving due east stopping for night vic XT963437. Unit moved SW at 150700 May and at 171245 May FOB was notified that unit needed water trailer to meet them at 1600 hrs vic YT938457. Unit returned to FOB from water re-supply point. (See tab N for further operations by 1st Mike Force Company).

(4) 3d Mike Force Company: The company received the mission of conducting a recon in force to determine a possible VC buildup in RZ DELTA. Unit was moved by truck at 161020 May to vic XT993525. At 161315 May vic XT999542, 4-6 VC were observed wearing black uniforms. No weapons were observed due to distance. At 161340 May two more VC were observed vic YT004547 wearing mixed uniforms, no weapons observed. Unit went into defensive position at 161800 May vic YT008568 and manned an ambush postion along LTL 1A. Negative sightings along route. Company was picked up by truck vic XT994521 at 171600 May and returned to FOB.

5. ENEMY INFORMATION:

a. On 3 May 67 MGF 957 contacted a multi-battalion sized force vic XT951669. This unit was equipped with AK-47's, RPG-2 rocket launchers, 2-3 heavy MG's, unknown type mortars and recoilless rifles. There were many ethnic Chinese fighting with this unit. The ethnic Chinese were all wearing khaki uniforms. One other group of VC were wearing black uniforms with camouflage soft hats. Some VC were wearing blue uniforms. This unit was well trained. They employed both fire and maneuver (well executed) and human wave attacks were attempted after many of the MGF were casualties. This unit is believed to have been 271 VC Main Force Regiment AKAQ761-subordinate to the 9th VC Division. This is based on reports placing 271 MF Regt in the immediate vic, and on a report that the 271 employs many ethnic Chinese mercenaries. On 11 May vic YT039408 recon mission 8 observed a total of 200-250 VC moving south from 0355H to 0515H. The VC were using flashlights and traveling in 4 different groups (approx company size). At 0734H the SIGMA FAC located 50-75 of the VC in an open area vic YT023378 and at 0802H TAC air was put in on the VC. Reaction by 2 platoons of 1st Company resulted in capturing 2 PW's and numerous documents that identified the units as 2d Company, 2d Bn, 273 VC Regt, AKAQ763, subordinate to VC 9th Div. Also, C-23, Medical Detachment, subordinate to 273 VC regt. Uniforms were mixed cloths of black, blue, and green. On 17 May MGF 957 located a large concentration of VC vic YT143578 (center of mass). There were many well used trails running coast-west throughout the area. Commo wire was run along some of the trails. All VC in the area were equipped with new CHICOM series of weapons (1956). Uniforms were well cared for and troops appeared well groomed and healthy. This is believed to have been part of 9th Div HQ Security. SPAR reports indicated that 9th VC Div is located northeast of center of mass coordinates given.

b. The 9th VC div is apparently massing its subordinate regiments in War Zone "D." This could be a temporary move for attacks on targets near War Zone "D" but more probably the 9th VC Div has moved into War Zone "D" on a permanent basis. Presently, the only two 9th Div Regiments believed to be in War Zone "D" are the 271 and 273 VC Regts. Both located and identified by Operation Blackjack 33.

Appendix G

Chronology of Logistics Events

June 1970	S-4 conference held in 5th Special Forces Group headquarters. The major topics discussed were camp conversion and proper procedures for submission of reports to the Logistical Support Center, 5th Special Forces Group.
July 1970	During this month, turn-in procedures were established with U.S. Army Support Command, Cam Ranh Bay, which allowed 5th Special Forces Group to dispose of unserviceable but economically repairable equipment through regular Army channels. Prior to this arrangement the group could only repair an item of equipment or cannibalize it. The results of this new arrangement substantially reduced the backlog in the group's maintenance facility and left key technicians free to perform other needed tasks in anticipation of the phase-down.
August 1970	Decreasing operational requirements caused the subtraction of three of the 5th Special Forces Group's dedicated C-7A aircraft. All airboats of the 5th Special Forces Group were reported to U.S. Military Assistance Command, Vietnam, as excess.
October 1970	Increased emphasis was placed on identifying and reporting equipment no longer required in anticipation of redeloyment. To facilitate the large future utilization, reporting procedures were revised through co-ordination with Military Assistance Command to permit full use of the data processing system at the U.S. Army depot, Cam Ranh Bay. This system provided rapid screening against United States Army, Vietnam, Military Assistance Service Funded, and Republic of Vietnam Armed Forces requirements. During October, 6,200 line items

were reported excess to operational requirements.

Two more C–7A dedicated aircraft were subtracted because of phase-down in operational requirements.

November 1970	Inventories were conducted in the group Logistical Support Center for reporting excesses in accordance with Department of the Army and Military Assistance Command procedures.
December 1970	Forward supply points located at Da Nang and Can Tho were closed. All excesses were backloaded to Nha Trang. During the month, 2,282 line items were laterally transferred in Vietnam.
	Turn-in of equipment began.
January 1971	Consolidated listings of balances reflected on stock record cards were submitted to United States Army Support Command, Cam Ranh Bay, for key punch action in providing disposition instructions. Operation Plan 183–70 was prepared. Forward supply point at Pleiku was closed. Equipment not transferred was convoyed to Nha Trang.
February 1971	Auditors were briefed on procedures being employed for closing out supply records for all accountable items. Stock record cards and related documents were reviewed, and auditors assured that current supply procedures were being followed. Auditors stated supply records would be destroyed in accordance with existing regulations and a certificate of destruction would be initiated to substantiate the destruction.
	Disposition of S–4 files was made.

Glossary

AA	Antiaircraft
ABN	Airborne
AK47	Chinese-made rifle
AO	Area of operations
ARVN	Army of the Republic of Vietnam
ATTLEBORO	Special operation in fall 1966 in III Corps
AW	Automatic weapons
BC	Body count
Berm	Dike or ledge
BLACKJACK 33	Special operation 27 April–24 May 1967 in III Corps in conjunction with Project Sigma, Detachment B–56
BLACKJACK 41	Special operation in spring 1967 in Seven Mountains region
CA	Civic action or civil affairs
CARE	Co-operative for American Relief Everywhere
C&C	Command and control
CIA	Captured in action
CIDG	Civilian Irregular Defense Group
COMUSMACV	Commander, United States Military Assistance Command, Vietnam
CP	Command post
CTZ	Corps tactical zone
DCO	Deputy commanding officer
DOW	Died of wounds
EM	Enlisted men
FAC	Forward air controller
FFORCEV	Field Force, Vietnam
FOB	Forward operations base
FWMAF	Free World Military Assistance Forces

G-2	Assistant chief of staff for military intelligence at an army, corps (field force in Vietnam), or division headquarters
GSW	Gun shot wound
GVN	Government of the Republic of Vietnam
IG	Inspector general
J-2	Assistant chief of staff for military intelligence, MACV
JGS	(Vietnamese) Joint General Staff
KBA	Killed by air
KIA	Killed in action
KKK	Khymer Kampuchea Krom (underground Cambodian faction)
LLDB	Lac Luong Dac Biet (Vietnamese name for their Special Forces)
LN	Liaison
LZ	Landing zone
M16	Standard American rifle
MACV	Military Assistance Command, Vietnam
MEDEVAC	Medical evacuation
MAF	Marine Amphibious Force
MF	MIKE Force
MG	Machine gun
MGF	Mobile guerrilla force
MIA	Missing in action
MIKE or MSF	Mobile strike force
MR	Morning report
Nung	Tribal group of non-Indonesian stock originally from the highlands of North Vietnam who provided special units for South Vietnam's Army
NVA	North Vietnamese Army
OB	Operations base
Off	Officers
OP	Observation post
OPCON	Operating control
OPORD	Operational order

PARASOL–SWITCHBACK	1963 program supplying funds for use with the CIDG in Vietnam
PCS	Permanent change of station
PF	Popular Forces (Vietnamese)
Prov	Provisional
PSYOPS	Psychological operations
PW	Prisoner of war
Quad	Four heavy machine guns that traverse from a single pedestal and which are fired simultaneously by one gunner
RF	Regional Forces (Vietnamese)
RPG2	Chinese-made rocket launcher
RR	Recoilless rifle
RVN	Republic of Vietnam
RZ	Reconnaissance zone
S–2	Officer in charge of the military intelligence section of a brigade or smaller unit
S–3	Officer in charge of the operations and training section of a brigade or smaller unit
S–4	Logistics officer of a brigade or smaller unit
S–5	Civil affairs officer of a brigade or smaller unit
SA	Small arms
Seabees	Naval construction engineers
SF	Special Forces
SFG	Special Forces Group
SFGA	Special Forces Group (Airborne)
SOP	Standing operating procedure
SPARS	Significant problem areas reports
SWITCHBACK	Operation in which the Army began assuming responsibility for U.S. participation in CIDG program, November 1962–July 1963
TAC	Tactical
TDY	Temporary duty
TAOR	Tactical area of responsibility
TOE	Table of organization and equipment
USAF	United States Air Force
USARV	United States Army, Vietnam
USASF	United States Army Special Forces

USMC	United States Marine Corps
USOM	United States Operations Mission
USSF	United States Special Forces
VC	Viet Cong
VN	Vietnamese
VNN	Vietnamese Nationals
VNSF	Vietnamese Special Forces
WIA	Wounded in action

Index

Abrams, General Creighton W.: 160
Advisers. *See* Military Assistance Command, Vietnam; Special Forces, U.S. Army; U.S. Army.
Aerial reconnaissance: 143
Agency for International Development: 6
Air cushion vehicles: 162
Air Force, U.S.: 80, 97, 115, 141, 142, 161
Air Force, Vietnamese: 141
Air strikes: 93, 135, 140
Air support: 87, 121
Airboats: 97, 155, 162
Aircraft. *See also* Helicopters.
  A1E-type: 137
  AC-47: 116
  B-40: 111
  C-7A: 142
  C-47: 93, 130, 161
  C-130: 142, 161
  F-4: 95
  F-105: 94
Airlift: 132, 141
Airlift command: 161
Airmobile operations: 84
Ambushes: 34, 40, 53, 135, 145
An Loa Valley: 148
Apache Force: 90, 91
Area development centers: 12, 33, 34, 35, 46
Area development programs: 10, 14, 34, 35, 48, 52, 78
A Shau: 83, 92, 93, 109, 162
ATTLEBORO operation: 83, 118

Ban Don: 63, 64
Ban Me Thuot: 22, 43, 64, 126
Ba Xoai: 119, 155
Binh Dinh Province: 87, 93
BLACKJACK landing zone: 119
BLACKJACK operations: 41, 83, 118, 119, 147, 169
Bo Duc Subsector: 114
Bong Son: 93
Booby traps: 105, 135, 145
Boorman, Staff Sergeant J.: 114
Border control system: 152
Border Surveillance-Control Operating Concept: 46
Border surveillance program: 36, 49
  camps: 47, 102, 104
  mission: 44
  operations: 52
  units: 33
"Briefback": 142
Brillante, Staff Sergeant: 111
Bronson, Captain: 94, 95
Brown, Specialist 4: 94, 95
Bu Dinh Secret Zone: 128, 133
Bu Dop: 83, 113, 114, 115, 155
Buildup, U.S.: 15, 77, 79, 80, 83, 84, 85, 99, 168
Buon Brieng: 63, 64
Buon Ea Yang, Camp: 87
Buon Enao (village): 12, 14, 24, 25, 27
Buon Enao experiment: 25–28, 32, 34, 37, 42, 43. *See also* Montagnard program.
Buon Enao Strike Force: 42
Buon Ho: 28, 43
Buon Krong: 28
Buon Mi Ga: 63
Buon Sar Pa: 63, 64
Buon Sar Pa Mobile Strike Force: 63
Buon Tah: 28
Bu Prang: 43, 63, 64, 154

Cambodia: 126, 154
Cambodian border: 33, 46, 115, 133
Cambodian operation: 152, 154
Cambodians. *See* Minority groups.
Camps, Civilian Irregular Defense Group. *See also* Buon Enao experiment.
  closure: 156
  construction: 49, 106, 127
  conversion: 85, 96, 108, 122, 151, 152, 156, 158
  deployment: 102, 108
  establishment: 15, 46, 84
  fighting: 55, 96, 106, 107, 121, 165, 170
  floating: 96, 106, 107, 165, 170

Camps, Civilian Irregular Defense Group—Continued
overrun: 40
turnover: 41, 44, 102, 122, 126, 157
weaknesses: 54
Can Tho: 58, 115, 119
Cao Dai sect: 24, 29
Cao Lanh: 90
CARE: 60
Casualties: 93, 109, 110, 115, 116, 118, 122, 148, 154. *See also* Medical evacuation.
Catholic parish priests: 33
Catholic youth program: 34
Cau Song Be: 148
Central Highlands: 6, 12
Central Intelligence Agency: 6, 147
Chain of command: 45, 46, 164, 166
Chamberlain, Captain Craig R.: 111, 113
Chau Doc: 89, 126
Chi Lang: 119, 155
Civic Action Guide: 121
Civic action programs: 17, 18, 41, 59, 62, 85, 87, 98, 124, 155, 164, 170, 171
Civilian Irregular Defense Group (CIDG) program: 32, 42, 49. *See also* Camps, Civilian Irregular Defense Group.
conventionalization: 48, 78, 79
conversion: 41, 85, 157
effectiveness: 53, 83
growth: 34, 85, 86, 87, 96, 108
inception: 6, 7
offensive role: 77, 78, 79
Combat reconnaissance platoons: 90, 91, 96
Command and control: 86, 141, 168
Command and General Staff College: 6
Commanders, U.S.: 80, 85, 141
Commando Training Center, Nha Trang: 4
Concept of Special Forces operations plan (1967): 98
Con Thien: 83, 110, 111, 163
Co-operative self-help concept: 63
Corps tactical zones: 30
I: 34, 48, 49, 55, 58, 83, 84, 102, 110
II: 34, 54, 63, 83, 84, 85, 93, 102
III: 84, 85, 102, 109, 113, 118
IV: 88, 90, 102, 108, 118, 155
"Counterpart" relationship, U.S. and Vietnamese Special Forces: 7–8, 12, 16–17, 28, 30, 35, 52, 101–02, 124

Crazy Horse operation: 87

Dak Mil: 63
Dak Pek Camp: 154
Dak Seang: 154
Dak To: 40
Dalat: 126
Da Nang: 32, 56, 58
Darlac Province: 6, 12, 22, 28, 34, 87
Davis, Major Paris D.: 94, 95
Delta Company Strike Force: 155
Delta Project: 53, 54, 81, 83, 87, 90, 91, 97, 101, 121, 134, 137, 140, 165, 169
Department of the Army: 161, 164
Department of Defense: 6, 30
Department of State: 6
Detachments: 6, 17, 168. *See also* Units, U. S. Army Special Forces, Detachments.
A: 9, 26, 28, 30, 41, 46, 49, 61, 74, 88, 89, 92, 97, 99, 100, 108, 135
B: 9, 30, 45, 46, 49, 54, 58, 74, 88, 89, 99
C: 9, 74, 92, 116
"Devil's Brigade": 4
Diem, Ngo Dinh: 4
Diem Regime: 35
Dien Bien Phu: 3
Dies, Sergeant Ronald: 94
Divisions, U.S. *See* Units, U.S. Army, Divisions.
Dong Ba Thin: 49, 91
Dong Ba Thin Training Center: 120
Dong Bao Thuong: 20. *See also* Montagnards.
Dong Xoai: 128
Donlon, Captain Roger H. C.: 56
Don Phuc: 118
Duc Hue: 154
Duc Lap: 154

Ea Ana: 28
"Eagle Flight": 54
Eagle Scouts: 90, 91
Educational projects: 25
Eisenhower, Dwight D.: 4
Engineer detachments: 41. *See also* Units, U.S. Army Special Forces.
Escalation. *See* Buildup, U.S.

Field force. *See* Units, U.S. Army, Field Forces.
Fighting camps. *See* Camps, Civilian Irregular Defense Group.

"Fighting fathers" program: 33
Fishhook area: 129
Floating camps. *See* Camps, Civilian Irregular Defense Group.
Flood campaign plan: 97
Forward supply points: 58,59, 121
Frederick, Lieutenant Colonel Robert Tryon: 4
Free World Military Assistance Forces: 85, 96, 98, 99, 101, 102, 104, 121, 128, 134, 159, 169, 170
French Army: 3
French Indochina War: 20
French Union: 3
Funding: 57,97. *See also* PARASOL-SWITCHBACK program; Special Forces Group (Airborne), 5th.

Garrett, Captain Chester: 114
Geneva armistice: 3
Gibson, Staff Sergeant: 111
Gillespie, Captain Vernon P.: 64
Gomez, Sergeant 1st Class: 112
Greek-letter projects: 137. *See also* Delta Project; Omega Project; Sigma Project.
Green Berets: 6, 40, 79, 93
Gritz, Captain James: 147

Hamlet militia: 15, 30, 32, 37, 48, 49
Hanoi: 4
"Hardened" camps. *See* Camps, Civilian Irregular Defense Group, fighting.
Harkins, General Paul D.: 28, 160
Helicopter Pioneers: 160
Helicopters: 97, 142, 143, 146, 162. *See also* Aircraft.
and chain of command: 166
H-34: 93
UH-1D: 107
HENRY CLAY operation: 83
Hewgley, Lieutenant Colonel Clarence T.: 147
Hiep Hoa: 15,54
Hill 158: 112
Historical units: 97
Hoa Cam Training Center (Da Nang): 32
Hoa Hao sect: 24, 49. *See also* Minority groups.
Hon Quan, Camp: 129
"Hot" war: 126
Hue: 92

Infantry, U.S.: 84, 85. *See also* Units, U.S. Army; U.S. Army.
Infiltration (by CIDG): 135, 143, 144, 145. *See also* Viet Cong, infiltration.
Insurgent wars: 163
Intelligence: 16, 81, 83, 87, 97, 99–100, 135, 165

Jarai (tribe): 25, 28
John F. Kennedy Center for Military Assistance. *See* Special Warfare Center.
Joint Chiefs of Staff: 6
Joint U.S.-Vietnamese Special Forces command. *See* "Counterpart" relationship, U.S. and Vietnamese Special Forces; U.S. Army Special Forces–Vietnamese Special Forces command.

Katum: 155
Kennedy, John F.: 5, 6
Khanh Hoa Province: 22
Khe Sanh: 110, 126
Khe Tre: 55,56
Kien Phong Province: 90, 155
Kien Tuong Province: 89
Kiet, Mr.: 111
King, Captain: 112
Kontum (city): 96, 126
Kontum Province: 34
Korean War: 163

Lac Luong Dac Biet: 111. *See also* Vietnamese Special Forces.
Lac Tien: 28
Lafayette Escadrille: 160
Land development program: 25. *See also* Area development programs.
Language problems: 17, 79, 164
Lang Vei, Camp: 83, 110, 126
Lansberry, Sergeant 1st Class: 112
Laos: 126
Laotian border: 33,46,92
Lao Tien: 43
Le Hai 21 operation: 87
Leadership: 52, 101
Leaping Lena Project: 53
Loc Ninh, Camp: 83, 127, 129, 130, 131
Loc Ninh District: 128

Loff, Sergeant 1st Class: 112
Logistic program: 134, 148, 158, 159
Logistic Support Center (Nha Trang): 45,58
Logistical support system: 17, 36,37, 57–58, 97, 98, 121, 122
Long-range reconnaissance teams: 53, 98, 134, 146, 165

McGuire rig: 147
Mai Loc: 154,156
Marine Corps, U.S.: 83, 84, 93, 110, 111, 112, 162, 163
Marines, Korean: 83
Marshall, General George C.: 3
Medical evacuation: 95, 105, 113, 115, 132, 147
Medical programs: 25, 59,63, 121
Military Assistance Command, Vietnam: 4, 12, 15, 45, 81, 85, 87, 90, 137, 151
  advisers: 88, 158
  civic action study: 62
  established: 28
  responsibilities: 35
  support to Special Forces: 160, 161
Military intelligence detachments: 97, 100
Mekong Delta: 12, 19, 34, 89, 102, 155, 162, 170
Mekong River: 107
MIKE Force: 140. *See also* Reaction forces.
Minh Thanh: 102
Minority groups: 19, 20, 24, 33, 49, 157, 158, 171, 172
Minutemen: 160
Missile, surface-to-air: 131
Mission support sites: 145
Mnong (tribe): 28
Mobile guerrilla forces: 8, 96, 98, 134, 135, 137, 148, 169
Mobile Strike Force Command, 5th: 116
Mobile strike forces: 54, 63, 84, 90, 127, 134, 137, 144, 145
  command and control: 141
  mission: 115, 116,117
  organization: 92, 116
  training: 96
Moc Hoa: 89
Monsoon operations: 162. *See also* Flood campaign plan.
Montagnard program: 24, 33, 34, 46
Montagnard uprising: 43, 45, 53, 63, 64, 167
Montagnards: 6, 12. *See also* Rhade (tribe).
  assimilation efforts: 157
  attitude toward Vietnamese: 9
  citizenship: 158
  Vietnam government, attitude toward: 19, 20
Morgan, Sergeant 1st Class David: 94
Motivation and indoctrination program: 152
Mountain commandos: 32
Mountain scouts: 15, 32, 37
Mountbatten, Vice Admiral Lord Louis: 3
My An, Camp: 109, 116
My Phuoc Tay, Camp: 109

Nam Dong: 55, 56
NATHAN HALE operation: 83
National Assembly, Vietnamese: 64
Navy, U.S.: 97, 108, 161, 162
Navy Construction Battalion: 41, 111. *See also* Seabees.
Nha Trang: 30, 45, 49, 119, 126, 159
Nha Trang Mobile Strike Force: 114, 116, 118
Night operations: 53, 97, 105, 165, 170
North Vietnamese Army: 78, 90, 92, 114, 115, 125, 126, 154, 163. *See also* Units, enemy.
Nui Giai Mountain: 118,119
Nung (tribe): 24, 55, 56, 57, 91. *See also* Units, mobile strike forces.

Okinawa: 4, 32, 57, 166
Omega Project: 81,97, 121, 134, 137, 169
OREGON Task Force: 84

Pacification program: 34,46, 64
Paramilitary forces: 19, 32, 33, 35, 161. *See also* Civilian Irregular Defense Group (CIDG) program; Popular Forces; Regional Forces.
PARASOL-SWITCHBACK program: 57
Peoples Republic of China: 4
Phan Thiet: 126
Phuoc Long Province: 113, 129
Phu Yen Province: 22

Pike, Geoffrey N.: 3
Plain of Reeds: 102
Plei Djereng: 107
Pleiku (city): 7, 58, 64, 96, 126
Pleiku Province: 34, 107
Plei Mrong, Camp: 40, 96, 102, 104
Plei Trap Valley: 161
PLOUGH plan: 3
Polei Krong: 40, 54
Popular Forces: 37, 63, 88, 89, 97, 172
Prek Klok, Camp: 109
Prisoner recovery operations: 148
Psychological operations: 17, 35, 41, 63, 98, 124, 164, 165, 171

Quang Duc Province: 22, 43
Quang Tri Province: 110
"Quick reacting supply and procurement procedures": 36
Qui Nhon: 126

Radio Research units: 97. *See also* Units, U.S. Army.
Rangers, U.S.: 160
Rangers, Vietnamese. *See* Vietnamese Rangers.
Rang Rang: 155
Reaction forces: 54, 80, 84, 87
Recondo School: 83, 96, 120, 121, 170
Reconnaissance teams: 91, 92, 143. *See also* Long-range reconnaissance teams.
Red ball system: 121, 122
Refugees: 19, 60, 145, 154
Regional Forces: 16, 83, 88, 89, 97, 98, 172
  conversion from CIDG: 85
  at Loc Ninh: 129–31
Reinburg, Sergeant 1st Class: 95
Republic of Vietnam Army: 4, 16, 35, 74, 83, 84, 97, 151, 159. *See also* Units, Army of the Republic of Vietnam; Vietnamese Rangers.
Republic of Vietnam government: 10, 24, 30, 35, 44
Resor, Stanley R.: 147
Resupply: 98, 121, 135, 137, 145. *See also* Logistical support system.
Revolutionary Development Support Program: 98, 124
Rhade (tribe): 6, 12, 20, 22, 23, 24, 64. *See also* Minority groups; Montagnards.
Rheault, Colonel Robert B.: 147–48
RIO BLANCO operation: 83
Roadrunner teams: 90, 137, 140, 143, 146
Rough Riders: 160

Safe areas, enemy: 140
Saigon: 30, 43, 45, 58
SAM HOUSTON operation: 83
Sampans: 155, 162
Sappers: 110, 111, 131, 154
Seabees: 112, 113
Seven Mountains: 83, 118, 148, 155
Sigma Project: 81, 83, 97, 121, 134, 137, 148, 169
Song Be (river): 129
Soui Da: 83
Soviet Union: 4
Special Forces, U.S. Army: 10, 35, 61, 151, 163, 169. *See also* Units, U.S. Army Special Forces.
  administration: 97, 98
  advisory effort: 8, 26, 35, 40, 87, 88, 101, 167, 168
  annual campaign: 97
  attitude of Vietnamese toward: 167
  command and control: 8, 17, 30, 147
  companies: 135
  definition: 9
  expansion: 32, 64, 85
  flexibility: 148
  future role: 173, 174
  lessons learned by: 171
  medical personnel: 7, 17
  missions: 16, 83, 84, 89, 145
  morale: 168
  motto: 175
  "navy": 170
  occupational specialties: 166
  organization: 171, 173
  training: 166
Special Forces (Provisional), Vietnam: 30, 40, 41, 42, 45, 46, 74
Special Forces, Vietnamese. *See* Vietnamese Special Forces.
Special Forces Group, 1st: 4, 166. *See also* Units, U.S. Army Special Forces.
Special Forces Group (Airborne), 5th: 10, 14, 30, 45, 64, 90, 102, 156, 157, 159, 166, 167. *See also* Units, U.S. Army Special Forces.
Special Forces Groups. *See* Units, U.S. Army Special Forces.

Special Forces handbooks: 98, 108, 170
Special Service Force, 1st: 3, 4
Special Warfare Center: 4, 6
"Spooky": 130, 132. *See also* Aircraft, C-47.
Stand-down: 155, 156, 157, 158, 159
Statistics: 165
Strategic hamlet program: 35, 42
Strike forces: 30, 34, 35, 84. *See also* Mobile guerrilla forces; Mobile strike forces.
  misuse of: 80, 141
  offensive role: 49-52
Suoi Da, Camp: 118
SWITCHBACK operation: 30, 35, 37, 40, 41, 42, 46, 57

Tactics: 10, 108, 160, 163, 164, 165
Tanks, North Vietnamese Army: 126
Task Force OREGON: 84
Tay Ninh Province: 62, 118, 148
Technical assistance teams. *See* Navy Construction Battalion: Seabees.
Tet offensive: 83, 122, 125, 126, 151
THAYER operation: 83
Thua Thien Province: 55
Thuong Duc: 154
Thuong Thoi, Camp: 83, 115
Tien Phuoc: 154
Tinh Bien: 119
Tong Le Chon, Camp: 109, 129, 155
Tra Cu: 154
Trailwatchers program: 33. *See also* Border surveillance program.
Training: 98, 166
  improvement of: 35, 104
  of strike forces: 37
  of trailwatchers: 26, 27, 33
  for unconventional operations: 140, 141
Trung Dung, Camp: 104

U Minh Forest: 148
Unconventional warfare: 6, 9, 134, 142, 143, 144, 145, 146, 147, 148. *See also* Civilian Irregular Defense Group (CIDG) program.
Units, Army of the Republic of Vietnam
  Battalion, 91st Ranger (Airborne): 90, 98, 137, 140
  Company, 883d Vietnamese Regional Forces: 93, 94, 95
Units, enemy
  Battalion, 512th Viet Cong: 118
  Division, 9th Viet Cong: 129
  Regiments
    12th North Vietnamese: 114
    28th North Vietnamese: 154
    40th North Vietnamese: 154
    60th North Vietnamese: 154
    84A North Vietnamese Artillery: 129, 130, 131
    141st North Vietnamese: 129, 131
    165th North Vietnamese: 131
    272d Viet Cong: 128, 131
    273d Viet Cong: 128, 129, 130, 131
    812th North Vietnamese: 110
Units, mobile strike forces
  Battalion, 3d Nung, III Corps: 114
  Companies
    1st, 3d Nung Battalion: 114
    D, IV Corps: 118, 119
  Corps
    III: 155
    IV: 116
  Buon Enao: 42
  Buon Sar Pa: 63
  Delta Company: 155
  Nha Trang: 114, 118
  Special Forces Group, 5th: 118, 119
Units, U.S. Army
  Brigade, 1st, 1st Infantry Division: 132
  Detachment, 403d Radio Research Special Operations: 100
  Divisions
    1st Cavalry (Airmobile): 83, 87
    1st Infantry: 127, 132, 133
    4th Infantry: 83
    101st Airborne: 83
  Field Forces
    I: 137
    II: 137
Units, U.S. Army Special Forces
  Companies
    A, 5th Special Forces Group: 154
    B, 5th Special Forces Group: 154
    C, 5th Special Forces Group: 154
  Detachments
    A-25: 115
    A-35: 25, 26
    A-101: 110
    A-110: 110
    A-251: 107
    A-322: 118

Units, U.S. Army Special Forces—Continued
Detachments—Continued
A-331: 127, 129, 130
A-425: 115
A-426: 116
A-430: 118
B-24: 96
B-41: 89
B-42: 87, 119
B-43: 90
B-55: 114, 118
B-56: 147
539th Engineer: 107
Force, 1st Special Service: 3, 4
Groups
1st: 4, 30
5th: 5, 74, 159
7th: 30
10th: 4
77th: 44
Team, Engineer, KB-7: 107
Units, U.S. Marine Corps, III Amphibious Force: 84, 93
U.S. Air Force. *See* Air Force, U.S.
U.S. Army: 58, 84, 98, 141, 153, 159, 160, 163, 169. *See also* Units, U.S. Army.
U.S. Army Special Forces. *See* Special Forces, U.S. Army.
U.S. Army Special Forces–Vietnamese Special Forces command: 54, 97, 135. *See also* "Counterpart" relationship, U.S. and Vietnamese Special Forces.

Vehicles, air cushion: 162
Viet Cong: 15, 19, 26, 81, 90. *See also* North Vietnamese Army; Units, enemy.
attacks: 55, 130
base areas: 108, 118, 135
infiltration: 54, 102, 110
infrastructure: 165
reaction to SWITCHBACK operation: 40, 41
shadow government: 53
training areas: 154
Viet Minh: 3, 4, 20
Vietnam Army: 4, 16, 35, 74, 83, 84, 97, 151, 159. *See also* Units, Army of the Republic of Vietnam; Vietnamese Rangers.
Vietnam Army Ranger Command: 156, 158
Vietnam Army Reconnaissance-Commando School: 121
Vietnam Independence League. *See* Viet Minh.
Vietnam War: 53, 78, 86, 151, 163, 164
Vietnamese Army Ranger Program: 157. *See also* Vietnamese Rangers.
Vietnamese Independence Day: 56
Vietnamese Joint General Staff–Military Assistance Command Planning Committee: 152
Vietnamese Rangers: 46, 54, 128, 152, 156, 157, 158. *See also* Units, Army of the Republic of Vietnam.
Vietnamese Special Forces: 4, 14, 26, 46, 90, 151. *See also* Leadership.
attitude toward civilians: 62
Civilian Irregular Defense Group program conversion: 158
detachments: 28, 101
organization: 101, 102
role: 16
view of Civilian Irregular Defense Group program: 40
Vietnamese Special Forces High Command: 74, 98, 105
Vietnamese Special Forces Training Center: 49, 91
Vietnamization: 96, 122, 151
Village defenders: 12, 26, 32, 34, 35. *See also* Hamlet militia.
Village defense system: 27
Village Self-Defense Program: 24, 26, 27, 28, 171
Vinh Gia, Camp: 102, 119, 155

Walter Reed psychiatric unit: 168
War Zones
C: 102, 109, 128, 148
D: 102, 129, 155
"Wars of liberation": 7
Waterborne operations: 165, 170. *See also* Flood campaign plan.
Waugh, Master Sergeant Billy: 94
Weapons Modernization Program: 127
Westmoreland, General William C.: 92, 97, 102, 120, 160
Williams, Sergeant 1st Class R.: 114

Zicaro, Sergeant: 111

☆ U.S. GOVERNMENT PRINTING OFFICE: 2004 306-180

Made in the USA
Monee, IL
04 November 2026

67385938R10134